Columbia University

Contributions to Education

Teachers College Series

No. 748

AMS PRESS
NEW YORK

A STUDY OF THOSE WHO INFLUENCE AND OF THOSE WHO ARE INFLUENCED IN DISCUSSION

By Ray Hamill Simpson

SUBMITTED IN PARTIAL FULFILLMENT OF THE REQUIREMENTS FOR THE DEGREE OF DOCTOR OF PHILOSOPHY IN THE FACULTY OF PHILOSOPHY, COLUMBIA UNIVERSITY

Published with the Approval of
Professor Percival M. Symonds, Sponsor

BUREAU OF PUBLICATIONS
TEACHERS COLLEGE, COLUMBIA UNIVERSITY
NEW YORK CITY
1938

Library of Congress Cataloging in Publication Data

Simpson, Ray Hamill, 1907-
A study of those who influence and of those who are influenced in discussion.

Reprint of the 1938 ed., issued in series: Teachers College, Columbia University. Contributions to education, no. 748.
Originally presented as the author's thesis, Columbia.
Bibliography: p.
1. Forums (Discussion and debate). 2. Social psychology. 3. Influence (Psychology). I. Title II. Series: Columbia University. Teachers College. Contributions to education, no. 748.
LC6519.S54 1972 301.11 70-177778
ISBN 0-404-55748-1

Reprinted by Special Arrangement with Teachers College Press, New York, New York

From the edition of 1938, New York
First AMS edition published in 1972
Manufactured in the United States

AMS PRESS, INC.
NEW YORK, N. Y. 10003

ACKNOWLEDGMENTS

THE WRITER'S INDEBTEDNESS is heavy to many who aided in making this study possible: To Professor Percival M. Symonds, sponsor of the study, for helpful guidance and advice; to Professor Helen Walker for suggestions on some of the statistical aspects of the problem; to Professor Lyman Bryson and Professor Wilbur Hallenbeck for theoretical and practical approaches to the problems of discussion; to Dr. Irving Lorge for very helpful suggestions on experimental technique as well as for making available some test materials; to Professor Harry Hollingworth for suggestions and encouragement; to Professor William McCall, Professor Goodwin Watson, and Professor Robert Thorndike for valuable criticisms; to Professor George Hartmann and Dr. Philip Eisenberg for making available some test materials; and to Dr. Anne Anastasi for aid in securing eleven control subjects. The writer is particularly grateful to all the students who so enthusiastically engaged in the discussions which form the basis for this study.

R. H. S.

CONTENTS

CONTENTS

CHAPTER I

PURPOSE AND SETTING OF THE STUDY

THE IMPORTANCE of the unsolved problems related to discussion is paralleled by the difficulties which face anyone interested in a scientific study of discussion. The present study represents primarily an attempt: first, to measure two types of influence which individuals exert in discussions with certain specified purposes; and second, to determine the relationship between certain individual traits and the two types of influence which have been measured. In order to orient the reader with regard to (1) the purposes involved in the discussions which were conducted as the basis for the study, (2) the particular types of individual influence measured, and therefore, (3) some of the qualifications which must necessarily be imposed upon the conclusions of the study, it seems desirable in this first section to outline very briefly the various purposes for which discussion may be conducted, and then to indicate with which of these purposes this study is related.

PURPOSES OF DISCUSSION. The reasons advanced for conducting particular discussions have been many and varied. Purposes of discussion from the standpoint of the individual, which may be proposed, some under certain circumstances and some under others, can be summarized briefly as follows:

1. To provide an opportunity for developing social attitudes, the ability to adjust to others in mutually enriching ways.
2. To provide an opportunity for developing new individual interests.
3. To broaden knowledge and acquire new skills. To enlarge the amount of subject matter which discussers have at their command.
4. To learn co-operative effort for a common end.

5. To give a recreative satisfying social experience.
6. To stabilize emotional life.
7. To clarify personal problems.
8. To give objectivity and tolerance, overcome misunderstanding, and conquer prejudice.
9. To emphasize the tentativeness of all knowledge.
10. To aid in the discovery of the unity of knowledge.
11. To gain group approval of a plan or project.
12. To negotiate.
13. To give an opportunity for intellectual struggle.
14. To test in the crucible of the thinking of others the opinions which one holds.
15. To give practice in acquiring the technique of conducting oneself intellectually, rather than emotionally, in the give and take, agreement and disagreement of discussion.
16. "To produce a collective thought by collective process."
17. To select from a number of alternative plans or solutions presented that plan or solution which the group will recommend as being most agreeable to it. This is the purpose with which this study is primarily concerned.

With regard to the purposes of discussion which have been enumerated, it should be noted that in some cases there is considerable overlap in purposes, and that as a rule no single purpose can be achieved without affecting in some way most of the other purposes mentioned. It should also be pointed out that the achieving of one purpose may not improve and may even hinder the group in the realization of another purpose. For instance, a group made up of individuals who are very competent as negotiators, and who therefore would presumably be efficient in that capacity when the purpose of the discussion is to negotiate, might have great difficulty in producing a "collective thought by collective processes." In other words, individual efficiency in negotiation or in "intellectual struggle" may create a mind-set in the discusser which will make it very difficult for him to work with a group to build up a common plan or procedure. Thus, simply because an individual or a group is found to be influential in the type of discussion involved in this study, we are not justified in assuming that the same individual would be most efficient when discussion is regarded from another point of view or when the purpose of the discussion is different.

Much attention has centered in recent years on the possibilities of discussion as a method "to produce collective thought by collective process," an expression used by Follett[8].* By this expression she means that the fundamental purpose of the discussion is to create a group idea, an idea which will be better than any one of the discusser's ideas alone and which will even be better than all the individuals' ideas added together. The type of discussion upon which the present study is based should be distinguished from the type spoken of by Follett. There is reason for thinking that discussion to produce a collective thought by collective process is more advanced as well as harder to achieve than is discussion based largely upon the convincing and persuasive ability of discussers who have the task of helping the group select one alternative from a number which are presented. The former more nearly represents an ideal type of discussion, while the latter represents discussion as it is frequently carried on today. With regard to the difference in purpose which has been mentioned, Leigh[10] in his book, *Group Leadership,* says, "A great deal of discussion represents, in reality, adoption by a group of the analysis made in advance by one of its members." Follett in her stimulating book on group organization points out: "I am sometimes told that mine is a counsel of perfection only to be realized in the millennium, but we cannot take the first step until we have chosen our path." It should be clearly recognized, then, that this is *not* a study of the efficiency of the group in producing a collective thought by collective process—except in so far as some degree of unanimity of opinion is arrived at with regard to the solution chosen by the group after discussion—but is rather an attempt to discover the characteristics which tend to be related to influence in *discussions as they are very frequently carried on today.*

More specifically, the types of influence with which we are concerned in this study are directly tied up with purpose listed as No.

* Numbers in brackets refer to items in Bibliography.

17: To select from a number of alternative plans or solutions presented that plan or solution which the group will recommend as being most agreeable to it. In general, the two types of individual influence measured are indicative of: first, the potency of the individual in getting the group to adopt his points of view as the best solutions to the problems discussed; and second, the efficiency of the individual in getting others to accept more or less permanently his points of view or something similar to them as the most desirable of the alternative solutions offered for consideration. A detailed account of the methods by which these measures of influence are obtained will be given later. The influence scores are then used to aid in answering such questions as the following:

Are individuals who are relatively more "intelligent" than others in the group also more influential in that group?

Do pencil and paper tests of dominance give any indication of dominance in a specific situation, such as that presented by the discussion situation?

Are the self-sufficient as measured by personality tests most influential in discussion?

Are radical members of discussion groups most influential?

Is religious background a factor of importance in determining the influence which an individual will have?

Are students who are best liked by classmates most influential?

Do individuals with artistic judgments most similar to those of experts tend to have most influence in discussion?

What types of individuals are most influenced by discussion?

TYPICAL BEHAVIOR PATTERNS IN DISCUSSION. What typically goes on within a discussion group under the circumstances involved in this experiment, when the purpose of the discussion is to select from a number of alternatives suggested that alternative which would be most acceptable to the group? The following brief and rather arbitrary analysis of characteristic comments and patterns within a particular discussion is given in order to indicate what usually happens within a discussion. In addition to giving a picture of the character of the give and take of discussion, this description should aid the reader in understanding the "influence scales" which have been used to indicate influence and which will be further described in a subsequent chapter. Without describing a discussion in detail, it will be noted that during the discussion of the type of problems in which we are here interested, there usually is something similar to the following sequence of events:

1. The lead is taken by one or more in the group and the general procedure is either verbally or tacitly agreed upon.

2. The group selects and formulates a problem, if this has not already been done, as was the case in the experimental discussions, and there is general acceptance of the problem for discussion.

3. All members of the group consider the problem which has been set, and the first person to comment on the problem is likely to do one of the following things:

a. Ask a question concerning the meaning of the problem or concerning the meaning of a particular word or phrase used in stating the problem.

b. State that a particular position is thoroughly untenable—the statement may or may not be accompanied by reasons.

c. Say that he supports such and such a position—again, reasons may or may not be given.

d. Ask someone in the group for his opinion on the problem as a whole or in a particular phase of the problem.

e. Make a comment concerning the purpose of this particular discussion and the time limit given, or ask a question concerning the type of decision it is desirable to reach in the course of this particular discussion.

4. The first comment bearing directly on the subject begins what may be termed the preliminary crystallization of opinions by the members of the group as well as the publicizing of the approximate place where this crystallized position occurs for each individual. For purposes of illustration let us assume that a particular group is discussing the question, "Should this college place more or less emphasis on what is commonly termed 'practical education'?" Or, stated in other words, "Should this college give more attention to courses and educational training which will prepare its students so that on being graduated they will be

better able to make a living?" If the alternatives which might be given as answers to this question be represented on a linear scale, then early in the discussion there is a crystallization of opinion, a taking of sides; and, if there are four in the group, the individual members of the group may be thought of as placing themselves on the decision scale in the following relative positions:

A B	D	C
Utilitarian emphasis favored		Cultural emphasis favored

It is evident that individuals A and B see almost eye to eye on this particular problem, although their reasons for doing so may be very different; whereas C takes almost exactly the opposite point of view—favoring strongly the cultural emphasis in college training. Student D apparently believes that neither the practical nor the cultural phase should be emphasized to the exclusion of the other, but that an intermediate position should be adopted. Obviously, with such a setup as this to begin with in a particular group, there is need for much marshaling of evidence and explanation of viewpoint if a group decision is to be reached. While we may expect the individuals through discussion to see the point of view of others more adequately, in this type of discussion the end result tends to be partially a compromise with some individuals giving in more than others.

From the preceding paragraph it might appear that marshaled evidence, well substantiated, is the only or at least the primary factor in producing or altering an opinion. In many cases this is rather far from the truth. As James Harvey Robinson has pointed out in his book, *The Mind in the Making*, many of our conclusions are reached before the mental structure which gives them reasonableness to us is built up. This "wish-thought" or rationalization process is the method by which in many cases our desires "determine in advance the outcome of the logical process." Because of emotional preferences or a previously built-up system of values which frequently are not consciously realized, we reach and hold tenaciously conclusions which with another system of values we would consider thoroughly untenable. For instance, if our whole philosophy is against any thoroughgoing revision of the capitalist system, we tend to discount about 99 per cent any suggestion regarding unemployment which might involve the survival of the capitalistic system itself. On the other hand, if we believe some form of communism or socialism is absolutely essential to get to the root of the trouble, we are very likely to minimize the value of proposals of those whose aim it is to bolster up the capitalistic system. Mind-sets, underlying systems of values, and long-time goals considered important are of paramount importance in determining the point of view an individual will have on a particular issue. This is particularly true if the individual tends to make his conclusions and do his thinking afterward.

Not only are the long-time goals very instrumental in producing a point of view, but the extent to which a particular issue is related to the achievement of one or more of the goals largely determines the amount of energy an individual will exert in pursuing the issue. More specifically with respect to discussion we may state that the energy shown by an individual in the discussion of a particular problem will depend to a considerable degree on the importance *to him* of the solution of the problem; that is, whether he thinks the solution will affect him in one way or another. The points made in this and the preceding paragraph indicate the importance of getting problems of some fundamental interest to the discussers. Specifically, in this study where the subjects were women college students, not all the problems selected were of major interest to all the students participating; however, the problems used were selected from a wide variety of fields so that all would find some problems of major interest, and many local school problems which students and college staff were considering at the time of the discussions were included. Measures were taken to meet the problems of interest and motivation, but it must be recognized that, as with all tests, interest in the subjects involved is an important variable which can never be entirely controlled.

Very early in discussions similar to those involved in this study there is a polarizing of individuals, a taking of sides. It is thus that opposing points of view are made apparent and some of the reasons back of divergent points of view brought out.

5. The second and subsequent opinions expressed will all be colored by preceding statements and by the personalities of the members of the group. Each comment will tend to fall into one of the following categories:

a. A rebuttal of the preceding point or points.
b. The introduction of new information or evidence.
c. The reiteration of an opinion previously made by someone else in the group.
d. The admission of a changed point of view or shift of opinion.
e. The recognition of evidence presented by a member of the group, but accompanied by a different conclusion from that reached by the person who first used the evidence. The difference in conclusion in many cases is caused by a different underlying system of values in the two persons involved.
f. A question concerning the accuracy or meaning of a particular bit of evidence.
g. An expression of a desire to crystallize the group decision and conclude the discussion of the particular problem involved.
h. A citing of the opinion of an expert.
i. The citing of a particular case favoring the viewpoint of the speaker. (A frequently used but dangerous practice since one case offers practically no evidence.)

6. Finally, an individual urges that the group accept a point of view which he favors and eventually his or some other point of view is adopted by the group.

Summary. In this study we are concerned with influence in discussion situations in which the purpose of the group is to select from possible solutions to a problem that solution which seems to the group to be best or most acceptable. It should be emphasized particularly that the group was not faced with the problem of building up a new solution to each problem, but that the task of the group was to decide which of the alternatives presented seems to have the most in its favor.

The purpose of the investigation with regard to influence in discussions of the type specified was to:

1. Test the ability of an individual:
   a. In getting the group to adopt his point of view and record it or something similar to it, as the decision of the group. This is termed influence of an immediate group nature, *or immediate group influence.*
   b. In basically changing the viewpoint of other members of the group. This may be called influence of a persisting sort, or *persisting individual influence.*
2. Correlate each type of influence with individual characteristics.

CHAPTER II

MEASURING INFLUENCE IN DISCUSSION

SINCE THE PURPOSE of the study was, briefly stated, to ascertain personality traits and characteristics which are related to amount of influence in discussion, and those which are related to being influenced by discussion, these three divisions in the procedure used in making the investigation stand out:

1. Determining for each individual the scores representing the amount of influence she exerted of the type previously described in each of a series of discussions, and determining the extent to which each individual was influenced by the discussion.
2. Selecting, giving, and scoring a series of personality tests in order to gain a knowledge of the characteristics of each individual. (Personality is here used in the broad sense, denoting all traits and characteristics which differentiate the individual.)
3. Correlating scores representing personality measurements and background characteristics (1) with scores representing amount and kind of influence in discussion, and (2) with scores representing the extent to which individuals were influenced by discussion.

SUBJECTS. The subjects used in this experiment were 185 women college students, each of whom* was attending a class in either Introductory Psychology or Educational Psychology at the time of experimentation. Of these students 161 served as experimental subjects and 24 as control subjects for the phase of the experiment in which control subjects were deemed necessary. Each of the 161 experimental subjects, with a few exceptions due to absences, engaged in four fifty-minute discussions. Also,

* With four exceptions.

each of the experimental subjects took a series of personality tests, which aggregated approximately nine hours for each student. Thus, the results of this study may be seen to rest upon about 1,400 hours of testing in addition to more than 500 hours of controlled and tested discussion.

INFLUENCE SCORES. The experiment was set up so that the two major types of influence, (1) immediate group influence or influence on the group decision, and (2) persisting individual influence, or influence on individuals which was retained for a time, could both be measured. The technique used in measuring the extent to which an individual was influential in formulating the group decision indicated immediately after the discussion of a particular problem, or immediate group influence, will first be described. Later a description will be given of the method used in measuring persisting individual influence.

MEASURING IMMEDIATE GROUP INFLUENCE. Eight series of problems were used as the subject matter for the experiment. Of the eight series seven were constructed by the writer. The eighth series (hereafter called Series III because it was the third series employed) was the McAdory Art Test. [26] Since the scoring method employed in the series of problems constructed by the writer is somewhat simpler than that used in the Art Test, the former method will be explained first.

CONSTRUCTING AND SCALING THE SOCIAL, ECONOMIC, POLITICAL, AND EDUCATIONAL QUESTIONS. For each problem used for discussion, except those of Series III, the writer took a current social, economic, political, or educational issue of a controversial nature. The issue of the problem was very briefly presented to an individual student who was then asked to indicate the stand, in answer to the question raised, which she considered to be most appropriate. To illustrate the method used in setting up the controversial issues so that each individual might indicate her position on the issues, the following problem, with the directions given first, will be utilized.

DIRECTIONS: Following these directions are seven problems upon each of which you are to indicate your position using the scale immediately after the problem. In each problem Alternative B, in general, represents the policy of maintaining the status quo, Alternative A represents a trend in one direction and Alternative C indicates a trend in almost the opposite direction. There are, of course, other possible alternatives, but you are asked to consider only these suggested in relation to each other. If you favor Alternative A, place a short vertical line on the scale at the extreme left of the scale. If you favor Alternative A with some modifications in the direction of an intermediate position, you should mark some place on the scale between the left end and the middle, the exact point being determined by the distance in the direction of Alternative B which you wish to represent yourself as standing on this particular problem. If you favor alternative C, place your vertical mark on the right-hand end of the scale. Locate your mark for each problem at the point on the scale which most nearly represents your present position with regard to that problem. Ask the instructor concerning anything about which you are not clear.

Problem I: *Gainful employment for married women.* In recent years much discussion has arisen concerning the advisability of encouraging women to seek gainful employment outside the home after marriage. Some say such a policy promotes individual happiness and that society also gains. The opposite is true according to others.

Alternative A. Women should be encouraged to look forward to gainful employment outside the home after marriage.

Alternative B. Our policy should be neither to encourage nor to discourage the practice of women working outside the home after marriage.

Alternative C. Women should not be encouraged to look forward to gainful employment outside the home after marriage. In fact, such work should be discouraged.

Alternative A	B	Alternative C

The above problem is the first from Series IV.* On the following pages are reproduced a copy of each of the seven series of questions which were constructed for the purpose of measuring discussion influence. Series I, II, and IV, together with the McAdory Art Test which constituted Series III, were all considered by four classes, two in Introductory Psychology and one in Educational Psychology, which the author taught at Barnard College during the first semester of the school year, 1936-1937, and one

* The scale used was 7¾ inches long.

in Introductory Psychology held during the second semester of the same year. Series V, VI, VII, and VIII were considered by a class in Educational Psychology during the course of the second semester of the same year.

Series of problems numbered I, II, IV, V, VI, VII, and VIII follow. Directions, given above, have been omitted.

SERIES I

(All questions in this series referred to the student's own college.)

Problem 1. Number of times classes meet per week.

Alternative A. Classes should meet more frequently.
Alternative B. Classes meet about the right number of times now.
Alternative C. Classes should meet less frequently than at present.

2. Amount of reading and study required.
   A. More should be required.
   B. Either more or less would be inadvisable.
   C. Less should be required.

3. Choice of work in a particular course.
   A. More choice should be left to the student as to which of various kinds of work in a course she shall do.
   B. About the same amount of choice as at present is wise.
   C. Less choice than is given at the present time is desirable.

4. Electives.
   A. More electives should be permitted.
   B. About the same number of electives as students now have is satisfactory.
   C. There should be more required courses.

5. Freedom of student action.
   A. More freedom should be given to the students in the choice of their activities and publications.
   B. The present situation is all right.
   C. Less freedom should be given to students than they have at the present time.

6. Emphasis on culture or making a living.
   A. More attention should be given to the preparation of the student for making a living.
   B. About the same distribution of emphasis as now exists is desirable.
   C. More attention should be given to the cultural side of the students' lives, even though this means decreasing the practical emphasis.

7. Emphasis on speed or accuracy.
   A. Too little attention is placed on speed in school work.
   B. The situation in this regard is about right.
   C. Too much attention is given to speed in school work.

8. Co-operation vs. individual activity.
   A. Too little emphasis is placed on co-operation.
   B. About the right amount of attention is given to co-operative activity.
   C. Too much attention is given to group activity in view of the individualistic nature of our economic system.

9. Marks.
   A. More emphasis should be placed on marks in school work.
   B. About the right amount of attention is given to marks.
   C. Less emphasis should be placed on marks in school work.

10. Individual vs. group instruction.
   A. More individual instruction should be given.
   B. About the same amount of individual and group instruction should be maintained.
   C. More group instruction should be given since individual instruction is much more expensive.

11. Tests.
   A. More tests should be given.
   B. The number of tests is all right.
   C. Fewer tests should be given.

SERIES II

Problem 1. The policy should be supported of having:

Alternative A. Higher institutions of learning entirely or at least largely supported by private individuals or organizations.

Alternative B. Higher institutions of learning jointly supported by the government or governments and private individuals or organizations.

Alternative C. Higher institutions of learning largely or entirely supported by the government or governments.

2. Ownership and control of the means of production and distribution.
   A. The government should own the means of production and distribution in order to aid the greatest number.
   B. The government should exercise a very large amount of control of the means of production and distribution but should not own the tools of production and distribution.

C. The government should adopt the laissez faire policy, and keep its hands off business almost entirely.

3. As our economic life is set up, some men work for a salary, some get paid directly according to the amount of work done, and some get paid a certain specified amount with a bonus for increased efficiency in work. We should encourage a policy of:

A. Paying everybody a fixed salary (not the same amount).

B. Paying a certain basic amount to everybody with a bonus for work done above a certain specified amount.

C. Paying everybody directly according to the amount of work he does.

4. Some say that today our people, generally speaking, are not given a broad enough education in our schools; others say that our people are not trained to do well enough any specific thing. Our schools should endeavor to:

A. Give a more broad and more general training (somewhat sacrificing efficiency in a specific field if that is necessary).

B. Maintain the present policy with regard to extensive vs. intensive training.

C. Make each individual intensively well trained in one specific field (somewhat sacrificing the broad general training if that is necessary).

SERIES IV

Problem 1. Gainful employment for married women. In recent years much discussion has arisen concerning the advisability of encouraging women to seek gainful employment outside the home after marriage. Some say such a policy promotes individual happiness and that society also gains. Others say that the opposite is true.

Alternative A. Women should be encouraged to look forward to gainful employment outside the home after marriage.

Alternative B. The tendency for women to work outside the home after marriage should be neither encouraged nor discouraged.

Alternative C. Women should not be encouraged to look forward to gainful employment outside the home after marriage. In fact, such work should be discouraged.

2. A smaller number of subjects should be studied at one time in college. Some colleges are today instituting the policy of having students study one or two subjects intensively for a six weeks period and then devoting all their time to another subject during the next six weeks, and so on through the year. Other schools are enlarging the number of studies students study at one time.

A. A smaller number of subjects should be studied at one time in college (this college).
B. About the right number of subjects are now being studied by students here.
C. A larger number of subjects should be studied at one time in college.

3. Should college education be more practical today? Some say such is the case; others say there is too great emphasis put on the practical aspects of education today.
   A. College education should be made more practical and utilitarian than it now is.
   B. The situation as it now exists is all right.
   C. More emphasis should be placed on the non-utilitarian aspects of college education.

4. What should be the attitude of intelligent people toward divorce? Some are saying today that family happiness would be increased if divorce were looked on with more disfavor than it now is; others believe it would be wiser to "liberalize" our attitudes toward divorce.
   A. The idea that divorce is a social evil should be encouraged.
   B. The present attitude with regard to this question should be retained.
   C. The idea that divorce is frequently best should be encouraged.

5. How much faith should be placed in experts? It is argued by some that the great social, educational, and political issues which confront our country are so complex that the ordinary citizen does not have the ability, or at least does not have the time to gain enough information about them to make intelligent decisions and that, therefore, the problems should be placed in the hands of experts for them to decide. Others say this would place too much power in the hands of experts and would be dangerous to democracy.
   A. More issues and problems should be left to experts.
   B. The present system in this respect is satisfactory.
   C. A smaller number of issues and problems should be left in the hands of experts; more should be decided directly by the people through the ballot.

6. Isolation of the United States vs. internationalism. The threat of war is a grave menace today. In view of this fact, some say the United States should adopt a policy of isolation, cut herself off as much as possible from the rest of the world, and thus avoid danger. The other major solution offered is that the United States should adopt a policy of responsibility and co-operation with other nations, which would mean that we would take a hand in old-world plans, policies, and disputes.
   A. The United States should adopt a policy of isolation.

B. The United States should attempt to steer a middle course with regard to this issue.

C. The United States should attempt to maintain peace by adopting a policy of internationalism.

7. In our civilization today most people have more than six hours of leisure per day. What policy should be encouraged with regard to the utilization of leisure for the greatest happiness?

A. Educate people to spend most of their leisure time in social pursuits, pursuits in which they will talk and play with other people.

B. Do as is being done at the present time.

C. Educate people to spend most of their leisure time in individual pursuits.

SERIES V

Problem 1. To what extent should minority groups be represented on teaching staffs?

Alternative A. Minority groups (religious, economic, and political) should have greater representation on public school teaching staffs than they now have.

Alternative B. The present representation of minority groups on teaching staffs is satisfactory.

Alternative C. Since a tenet of democracy is majority control of institutions, minority groups should have a smaller representation on our teaching staffs than they have today.

2. Should we teach competition in our schools?

A. Pupils should be taught in our schools to be more adept in the use of present-day competitive business methods since such preparation will better prepare them for life as they will find it outside the school.

B. Present practice in this respect is satisfactory.

C. Pupils should be taught co-operation rather than competition.

3. Should the emphasis in education be on child life or on adult life?

A. Education should be considered more a preparation for adult life than it now is.

B. The present emphasis is all right.

C. Educators should aid the child in meeting his day-to-day problems rather than give so much attention to probable problems the child may face five, ten, or fifteen years from now.

4. Should our primary emphasis in education be on learning methods of thinking or on accumulating facts?

A. Educators should place more emphasis than they are at present doing on the techniques of thinking even though this means less emphasis on the facts or materials for thinking.

B. Present practice in this regard is satisfactory.

C. Educators should place more emphasis on the accumulation of facts, the materials for thinking, even though this means less emphasis on techniques and methods of thinking.

5. What should be the relation of the schools to the government? The government should:

A. Utilize the schools to change the present culture and produce a new social and economic order.

B. Continue its present policy.

C. Consider the educational system as an institution whose primary and almost exclusive function is to inculcate the best learning of the past in the minds of young people.

6. Who shall control the schools?

A. Since the government is largely controlled by a comparatively small number of people, the schools, if they are to be truly democratic, should steer clear of government control as far as possible.

B. The present amount of government control of schools is satisfactory.

C. The schools should be very directly controlled by local and/or state governments and/or the national government, since they are the governments' chief means of promoting the welfare of all citizens.

7. Assuming that there will inevitably be indoctrination, the schools should be used for:

A. Conservative indoctrination.

B. The present type of indoctrination.

C. Radical indoctrination.

SERIES VI

Problem 1. Should educators control the movies?

Alternative A. Educators should have direct control or at least censorship over the movies since the cinema is believed to be one of the most potent educational influences.

Alternative B. Educators should only attempt to exert the type of influence they are now exerting.

Alternative C. Educators should let movie producers and distributors be alone responsible for their products.

2. Should dialects be encouraged in our schools?

A. Our elementary and secondary schools should take more active steps to encourage a standard speech, thereby largely eliminating dialects in this country.

B. Present practice in this respect is satisfactory.

C. Our elementary and secondary schools should not attempt to eliminate dialects in this country since standardization of this type is not desirable.

3. What type of citizen is it most desirable to produce in our society?

A. Educators would do well to attempt to produce contented citizens by putting little or no emphasis on criticism and questioning on the part of most elementary and secondary school pupils.

B. The present emphasis on criticism and questioning is satisfactory.

C. Educators should attempt to make most elementary and secondary school pupils as critical of all plans, orders, and ideas as possible.

4. How inclusive is the job of a scientist? Today we have a struggle of importance to education between those who desire more "pure science" investigators and those who say all scientists should be interested in the application of their findings. From your point of view which alternative represents the most desirable view?

A. The purpose of science is to find facts, not to interpret them or decide how they shall be used. Therefore, a real scientist should not concern himself with determining the aims, objectives, or ideals of life or the applications of his findings as a part of his work as a scientist.

B. The present attitude of most scientists is satisfactory.

C. A real scientist selects problems which have practical importance and, after getting his results, is interested in having them applied.

5. Inasmuch as the schools must today function in a society in which groups struggle against each other, what should be educators' attitude toward controversial issues? Emphasis upon which of the following alternatives would, in your opinion, probably produce the happiest personalities?

A. Since educators in secondary schools are dealing with immature individuals, they should avoid discussion of controversial issues as much as possible.

B. The present emphasis is satisfactory.

C. Controversial issues should be given a great deal of attention in our secondary schools.

6. In the light of your knowledge of the types of problems which most secondary school pupils face now and will face later on, how much emphasis should be placed on their following the decisions of more "intelligent" individuals as opposed to thinking through their problems themselves?

A. Educators should teach most youngsters to seek and depend more on the advice and decisions of more intelligent and more mature people.

B. The present intermediate position is all right.

C. Teaching youngsters to think through their own problems and their own decisions should be emphasized.

SERIES VII

Problem 1. What racial attitudes should be encouraged in our schools? In many sections of our country the question of working out a desirable relationship between races, particularly between black and white, is very acute. Which of the following alternatives should be encouraged?

Alternative A. Segregate each race as much as possible and thus eliminate many of the relationship problems.

Alternative B. Do not encourage segregation of races but maintain barriers between the races.

Alternative C. Encourage the elimination of all barriers, including intermarriage between races.

2. Should marks in a particular course at Barnard College be based upon the final position of the individual in the subject matter of the course, in so far as that can be determined; or should they be based on the amount of progress the individuals made during the course?

A. Marks in a course should be based upon the amounts of knowledge which students have at the end of the course, regardless of the amount of knowledge each had at the beginning of the course.

B. A compromise between Alternative A and Alternative C is desirable.

C. Marks at the end of a college course should be based upon the relative amounts of gain in knowledge the students have made during the course. In other words, the person who made the greatest gain in knowledge, in so far as that can be measured by protests and retests, would get the highest mark.

3. Should teachers of social studies have a definite political philosophy?

A. Each teacher of social studies, conservative and liberal alike, should have a rather definite political philosophy and let his students know where he stands.

B. Each teacher of social studies should have a live political philosophy but should avoid letting his students know where he stands with regard to politics.

C. It is undesirable for a public school teacher of social studies to have a definite political philosophy since it is likely to bias his teaching.

4. Should marks in Barnard be based simply upon achievement or upon achievement related to the potential ability of the individual in the subject involved?

A. Marks should be based simply upon achievement. In other words the person who knows the most should get the highest mark.

B. A compromise between Alternative A and Alternative C is desirable.

C. Marks should be based upon achievement related to the potential ability of the individual involved. In other words, a person with an extremely high I. Q. would be expected to know more than a person with a relatively lower I. Q. to get the same mark.

5. Should we have teaching of morality in our schools?

A. Teaching of morality (not necessarily religion) should become an established part of the school curriculum.

B. Present practice with regard to the teaching of morality is satisfactory.

C. Teaching of morality should not be attempted by our schools, chiefly because the job of the schools is to promote intellectual development.

6. Should we indoctrinate for democracy?

A. Teachers in our public schools should indoctrinate for democracy and for democratic ideals.

B. A compromise between Alternatives A and C is desirable.

C. Teachers should scrupulously avoid indoctrinating for democracy, fascism, communism, or other political or economic ideologies.

7. Should individual differences for which the individual is not responsible cut down or increase the amount of reward which he gets for his labor? Our society should encourage:

A. Paying everybody entirely according to the amount produced, thereby emphasizing individual differences.

B. Reaching a satisfactory compromise between A and C.

C. Paying everybody who works conscientiously the same amount, thereby not penalizing the individual who has a lack of ability for which he was not responsible.

SERIES VIII

Problem 1. What should determine which teachers should be promoted in a particular school system?

Alternative A. Seniority should be the primary basis for promotion in public school systems. This means that the teachers who have been in the system the longest would have the highest ranks and get the best salaries.

Alternative B. A compromise between A and C is desirable.

Alternative C. Seniority should be considered of minor importance in considering teachers for promotion. This position assumes that some other more desirable criterion or criteria would be used in determining promotion.

2. Should public school teachers support teachers' unions?
   A. Teachers in public schools should support and encourage the development of strong teachers' unions which would fight for those things which teachers deem desirable.
   B. Cautious teacher groups should be supported by public school teachers.
   C. Since the teacher is a public servant, unique in his responsibilities for tender minds, he should not resort to the organizational techniques of people engaged in competitive business and industrial life. Teachers' unions should not be supported.

3. Should subject matter divisions be broken down in our elementary and secondary school systems?
   A. Subject matter divisions should be broken down as rapidly as possible in our elementary and secondary schools in order more adequately to integrate the educative process.
   B. A compromise between alternatives A and C is desirable.
   C. Subject matter divisions should be made sharper in order that a child may not be confused by having to consider too many phases of life at any particular moment.

4. What part, if any, should teachers play in social reconstruction?
   A. Since the work of the teacher is to educate the child in the way that the community wants him to be educated, the teacher should not align himself with radical movements interested in social reconstruction.
   B. A compromise between A and C is desirable.
   C. The teacher should play a very active role in organizations whose aim is to supplant features of the present organization of society with other features which seem to be more desirable.

5. Should college teachers be promoted primarily on the basis of classroom activities or on some other basis?
   A. College teachers should be promoted practically exclusively on the basis of the quality of their classroom teaching.
   B. A compromise between A and C is desirable.
   C. College teachers should be promoted on some other basis besides the quality of their classroom teaching.

6. How is artistic appreciation best taught?
   A. In developing artistic appreciation in our schools more attention should be paid to giving the students actual experience in producing art in those fields in which appreciation is to be built up.
   B. A compromise between A and C is desirable.
   C. The best way to produce appreciation is not through production or attempted production of art but through familiarity with the best

in art. Thus, more attention should be given to explanation and analysis of the great works of art.

7. How much weight should knowledge of subject matter be given in the selection and promotion of teachers?
   A. Knowledge of subject matter should be practically the exclusive criterion used in the selection and promotion of teachers.
   B. A compromise between A and C is desirable.
   C. Other factors besides knowledge of subject matter should be given the heaviest weight in the selection and promotion of teachers.

8. To what grade level should our schools be free?
   A. "Free schools" for young people beyond high school age are not justifiable.
   B. A compromise between A and C is desirable.
   C. Opportunity for free education for all who can profit from it, starting with the kindergarten and going on up through the university, should be given.

LIMITATIONS OF THE RESPONSE DEVICE. It has already been pointed out that the types of influence which the response device is designed to measure are directly related only to discussions in which the purpose is to select from a list of alternative solutions offered, that solution which the group will undertake to recommend as being most acceptable to it. The device probably has little or no significance in discussions in which the purpose is to build up in the group a *new* solution to the problem which has been presented.

The second major limitation is concerned with the restrictions imposed on the person responding. The person indicating her response was restricted to the two main responses and points intermediate to them. A person with another solution which might be at right angles to the linear scale presented for a particular problem could not respond in the way that she considered would be the most appropriate to the problem. Consequently, in the discussion of this particular problem, that individual would be at a disadvantage, since to none of the alternatives permitted by the setup of the problem could she give her wholehearted support. The situation is somewhat analogous to that sometimes encountered in multiple choice tests. For some of the problems none of the answers provided by the test seem to have much merit in the eyes of the person taking the test, since the answers which the student considers best have been left out. The obvious difficulty in permitting the student to give her own answer arises when an attempt is made to score unauthorized alternatives which most of the group did not consider. In the response device used with the present questions, answers not on the linear scale would have made almost impossible the relative localization of responses in such a way that the shift in group opinion could be determined and quantified. However, in only three or four of the problems used did students indicate a desire to give a solution to the problems which could not well be placed somewhere in the linear scale. Also, the problems were so presented that it was a matter of which of the alternatives presented was to be considered the best, rather than a question of what the best possible solution to the question might be.

The third of the major limitations pertaining to the response device rests on the fact that the student was not permitted to give the particular conditions under which a particular policy would, in her opinion, be most desirable. The permission to use such qualifications, however desirable from some standpoints, would have made the determining and recording of comparable responses an impossible task. It is conceivable that an individual might choose one alternative under certain conditions and quite a different alternative under other conditions, and in such cases the response device would have distinct limitations. If the purpose of the discussions had been to build a new plan in the group, this device would, of course, have been of little or no value. The seriousness of the major limitations is most difficult if not impossible to estimate in any objective fashion; however, they certainly should be borne in mind in considering results.

Presenting the problems. The problems were first presented to all students individually, a series at a time, about five days before the discussion of a particular series was to take place. The students were asked to indicate their individual positions on each problem, using the scales previously described which accompanied each series of problems, a scale for each problem. These pretest responses were recorded by the investigator who used a scale seven and three-quarters inches long made of 50 equal units, constructed for the purpose, which scale was placed directly beneath each response on each problem and the unit corresponding to the response noted. Thus, a response which indicated that the student agreed very closely with Alternative A in a problem would be recorded with a very low number, the exact number depending on how close to the left side of the scale the student had placed her vertical mark for that particular problem; a mark in the intermediate section of the scale would indicate that Alternative B was favored; and a comparatively high number would indicate that Alternative C was the most acceptable to the student doing the evaluating.

Organizing the groups. As has been indicated on the preceding pages, about five days before a discussion was to be conducted on a particular series of questions, each student was asked to indicate her position on each question in the series, using the scales previously described for all series except Series III, the McAdory Art Test, which was administered with the use of the standard response sheet made by the originator of the test. The results of this first set of responses, called the pretest, were then recorded for each individual and the groups for discussion were made up. Groups with four names in each were selected so that during the course of the semester, in so far as it was practical, no two individuals would be together in a group more than once. The purposes back of this latter measure were two. Before stating these two purposes let it be noted that in the first place the possibility of influencing others in discussion depends, in addition to the factor of the potential influencer's characteristics, upon the susceptibility of those in the group to influence in discussion; and in the second place upon the influencing caliber of the other potential influencers in the group. Thus, the specific purposes involved in having each student discuss the problems with as many other students as possible were: (1) To give each an opportunity to exert her influence on a good sampling of the class involved, and (2) to allow each to compete in influencing with a fair sampling of those who made up the class. To summarize, we may say that the possibilities for influencing as well as the quality of the opposition competing for these possibilities were largely equalized by the method of rotating individuals so that no one influenced or competed in influencing with any other person more often than was necessary. Additional considerations in the grouping of students for discussion of the McAdory Art Test will be described later.

THE OPTIMUM SIZE OF DISCUSSION GROUPS. The size of the groups, four in each (with a few exceptions), while rather arbitrarily decided, was largely based upon experience with student groups of various sizes before the experiment began. It was felt that groups which were much larger would not give each individual a chance to express herself as often, and that smaller groups would sometimes be handicapped by lack of sufficient ideas for real, live discussion. (The sizes of the classes made it necessary to put three students in a few groups, and five in others.) The writer believes there is no simple answer to the question, "What is the optimum size of a group gathered for discussion purposes?" In a particular situation the answer would seem to depend largely upon the relationship between three important factors: (1) The *time* allotted to the discussion of a specified subject; (2) the *purpose or purposes* of the discussion, and (3) the amount of *information not previously criticized* in the crucible of discussion which the members of the group have at their command. Much experimentation needs to be done before the following views can be entirely substantiated; however, the writer's experience leads him to accept them tentatively as representing likely hypotheses:

1. Other things being equal, the more *time* there is at the disposal of a group for the discussion of a particular problem, the larger the group can profitably be. Diagrammatically this may be illustrated by the following figure.

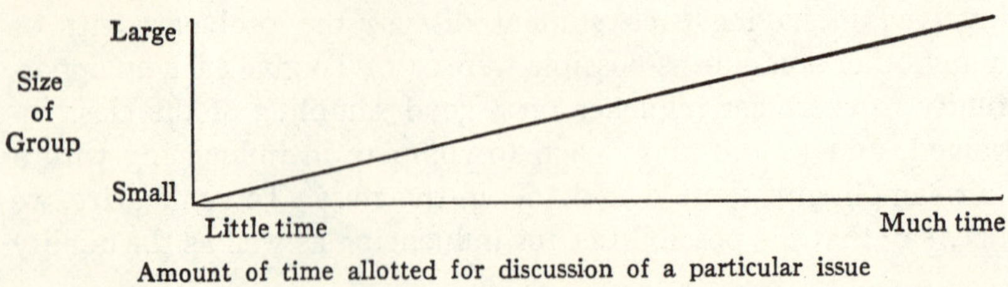

2. Other things being equal:

a. If the *purpose* of the group is to interchange a great amount and variety of information, then the group should be relatively large.

b. If the *purpose* is to have the ideas and information of each individual in the group thoroughly criticized and analyzed, then the group should be relatively small. The following illustration indicates the relationship of size to the items relating to purposes described in this and the preceding item (items *a* and *b*).

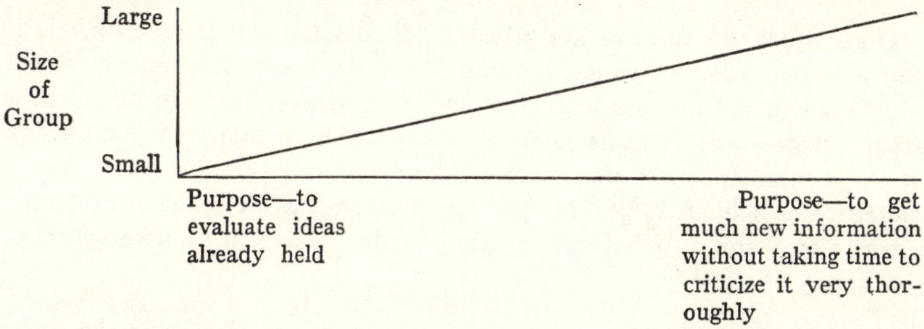

c. If the *purpose* of the discussion is to give each individual an opportunity to express himself, or to enjoy verbal fencing, then the size of the group should be relatively small.

d. If the *purpose* of the discussion is to give each individual in the group an opportunity to express himself, and this be combined with the purpose of reaching a group decision quickly, then again the size of the group should be relatively small.

3. Other things being equal, if the *amount of undiscussed information* which the members of the group possess is relatively small, then the group should be relatively large; conversely, if the amount of undiscussed information on the subject which is possessed by the group is relatively large, then the group should be relatively small. The following sketch illustrates the idea:

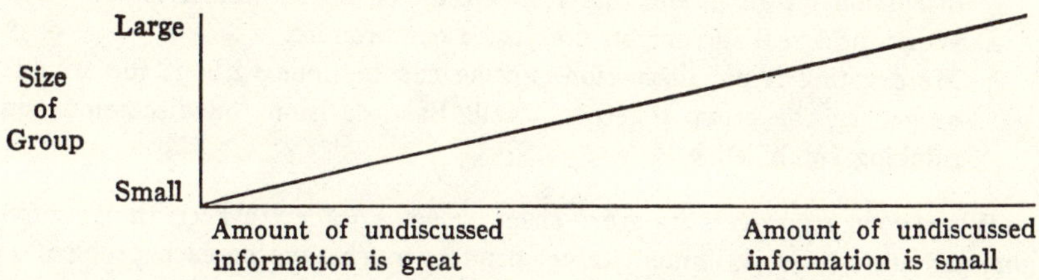

CONDUCTING THE DISCUSSIONS. On the day when the discussions were to be held, the students were told that the class would be divided into small groups for the discussion of some problems involving adjustment, motivation, and the elements of social psychology. The primary purpose of the discussion was, of course, not divulged. The names of the students who were to make up each group were announced, and each member was given a copy of the series of problems to be discussed along with the following on a sheet of directions:

DIRECTIONS: Today you are asked to discuss the problems upon which you indicated your individual positions some days ago. The opinion sheets which you turned in were inspected, and each of you has been placed in a group composed of students some of whom will not agree with you on any of the issues. An effort was made in forming the groups to get as much difference of opinion as possible in order that real, live discussion may take place in each group. Individual efficiency in this classroom exercise will consist largely of:

1. presenting clearly and adequately your own points of view on the problems involved;
2. listening carefully to the points of view presented by other members of the group and incorporating the worth-while parts which they contribute into your own thinking; and
3. aiding other members of the group in reaching a group decision, which is to be recorded on the group answer sheet, which decision should represent the best critical and deliberative thinking of which the group is capable.

Remember particularly these things:

1. The majority is frequently wrong.
2. If your point of view differs from that of someone else, the difference in viewpoint is caused by a difference in perspective, which can be largely eliminated by explanation, discussion, and criticism.
3. A person who has lost her temper has no place in discussion. Keep the discussion on an intellectual rather than on an emotional basis.
4. Work for a real agreement, not just a compromise.
5. Mere voting is not discussion. Voting can be done without the trouble of getting the group together at all. Base decisions on discussion and thinking combined.

When your group gets its score sheet, select a leader to keep things moving, and start to work immediately. Spend as much time on each problem as seems advisable to reach the group decision which is recorded on the group score sheet, finishing the discussion of all problems if there is sufficient time.

The discussions were all conducted in the classroom, its ample size making it possible to do this without the discussion of one group interfering with that of others. During the work the experimenter kept an eye on all groups, moving about the room to see that all were progressing in a satisfactory fashion. During the first discussions held in each class it was found that stalemates of various sorts occasionally occurred and had to be resolved. In two groups it was found that the cause of the difficulty in each case was a girl's losing her temper because others disagreed with her. No trouble of this sort was experienced after the first discussion. On several occasions the source of the trouble was found to be irreconcilable differences of opinion on a particular problem among members of the group. In each case only a bit of direction was needed to get the discussion moving again. In general, the girls enjoyed the discussions and entered into them with admirable zest.

EVALUATING IMMEDIATE GROUP INFLUENCE. Since the group decisions had been indicated on the same kinds of scales, one for each problem discussed by the group, as had the individual opinions made five days before the discussion took place, it was possible to use the same numbered scale for recording group decisions as had been used in recording individual decisions. This was done, and for each group we had tabulations on a particular problem which looked something like this:

Series I, Problem 1

Individuals in group	Position on pretest	Immediate group influence score on problem
Spragg	15	—14
Knott	12	—14
Elson	30	— 6
Jenson	47	14
Group mean on pretest = *26*		

GROUP DECISION ARRIVED AT AFTER DISCUSSION = 40

The scores under the heading "Position on pretest" indicate that both Spragg and Knott tended to favor Alternative A as opposed to Alternative C, and that they did not entirely favor Alternative A. Apparently Elson favors a definitely intermediate stand between the extremes A and C with a very slight tendency, it would appear, to favor C as opposed to A. Jenson at the beginning of the discussion is obviously in almost complete agreement with the alternative designated as C.

In the preceding paragraph the assumption has been made that if an individual supports any position in the discussion, it will probably be the same or a similar position to that which she indicated on the pretest. In other words, since both Spragg and Knott indicated on the pretest that they favored a position about halfway between Alternatives A and B, we assume that they will support similar positions in discussion. Two factors which would tend to make the discussion position differ from that on the pretest are the lapse of time and the different types of situation involved. From observation of the discussions and remarks made by students, the writer gained the impression that the pretest position was usually supported in discussion; however, it should be recognized that evidence on this point is largely subjective. In the experimental groups remarks were frequently prefaced by statements similar to, "When we were asked to select from these alternatives before, I selected this one because . . . ," indicating that a similar position was being maintained to that indicated on the pretest. Also, in the direction sheet given to each discusser, attention was called to the fact that an opinion had previously been expressed, and the individuals were instructed that one factor in the efficient progress of the discussion would be "presenting clearly and adequately your own points of view on the problems involved." The pretest viewpoint is used as that most likely to be maintained rather than an average of pretest and retest since the latter would seem to be partially a direct or indirect result of stands taken by others in the group. If more time had been given to the discussion of each problem, the dynamic aspect of an individual's position would undoubtedly have been more serious from this standpoint. However, even in the limited time given for discussion of a particular problem, it is quite possible that an individual's attitude changed, and that his expressed stand was modified. In making the basic assumption discussed in this paragraph the author realizes that in some cases the individual will get too much credit and in other cases not enough, with the exact net result of such variations impossible to determine.

By inspection of positions in Problem 1 in Series I it will be seen that Jenson was apparently the individual most influential in swinging the group and getting it to place the group decision far over in the direction of Alternative C—at least the weight of group opinion which was very close to the exact middle of the scale (26) at the beginning of the discussion shifted far (to 40) in the direction of the position which Jenson presumably argued for. For this reason the scoring should give Jenson a relatively high score for immediate group influence in this problem. On the other hand, Spragg and Knott apparently lost out in this particular phase of the discussion, in so far as influencing the group decision goes, and so should get a comparatively low score or a negative score. With regard to Elson it would seem that, since she occupied a position farther in the direction of Alternative C than the pretest mean, in the first marshaling of evidence she probably favored Jenson's position, at least as opposed to that of Spragg and Knott. However, since the group position moved over and quite some distance beyond her position, her score will be intermediate. The methods used in influencing were undoubtedly quite varied. No attempt was made in this study to investigate various methods used or their relative effectiveness.

Various scoring methods were tried out, but the one finally adopted gives a positive score on a problem to an individual if the group decision, 40 in this case, is closer to that individual's pretest score than was the group mean, 26 here, computed by adding the pretest scores of the individuals in the group and dividing by the number in the group. If the group decision is farther away from the individual's score than the mean of the pretest scores, then the individual's score is negative. The exact size of the individual's positive or negative score depends upon just how much nearer (for positive scores) or farther (for negative scores) the group decision is from the individual pretest opinion than is the pretest mean.

On a single problem in one series the positions of the students in the group composed of Spragg, Knott, Elson, and Jenson would look somewhat as follows on the scale:

		Pretest Mean "26"		Group Decision "40"	
12 Knott	15 Spragg		30 Elson		47 Jenson

This scale illustrates graphically the statement that was previously made to the effect that Knott and Spragg definitely lost ground so far as swinging the group decision went, since it was 14 points farther away from their pretest positions (the positions they presumably argued for) than was the pretest mean (which represents the balance of individual positions at the beginning of the discussion). As the numerical difference between 26 and 40, or 14, represents a loss for Knott and Spragg for this particular problem each of them is given a score of —14. By scoring in like fashion it will be seen that Jenson should be credited with 14 on this problem. In the case of Elson the group decision is six points farther from her pretest position than is the pretest mean, so she is given a score of —6.

It may be objected that Elson could only gain four points on this problem, which is quite true. However, investigation revealed that those who received the highest influence scores in the groups were, on the average, approximately the same distance from the group mean as were those who received the lowest scores for influence. This was concluded after making an intensive analysis of the influence scores and distances from group means (of those in the largest class used in the experiment) which were made on Series II and IV (representing a good sampling of the social, political, educational, and economic issues used) and Series III (representing esthetic problems). The results indicated that on Series III the average distance on all questions from the pretest group mean for the two lowest individuals from the standpoint of influence in each group was 46.6; while the average distance from the group mean of the two highest individuals from the standpoint of influence was 44.5. This would give a slight difference of 2.1 points favoring those who actually had the least influence; however, the difference is not statistically significant, being within the range that we might expect by chance. On the other hand, in Series II and IV there was a slight difference tending to favor the two highest in the group who had an average deviation from the pretest group mean of 39.8, while the lowest two had an average deviation of 37.3; again, a difference of this size might have been expected by chance. Combining the results (and giving three weights to the average results from Series II and IV since there were three series of this sort) we get a net difference of .5 favoring the two highest; this difference does not approach statistical significance and we conclude that high scores were not significantly related to great initial distance of individuals from the group mean.

A second possible objection which might reasonably be raised to this scoring method should also be mentioned. It is involved in a distribution of pretest positions such as the following:

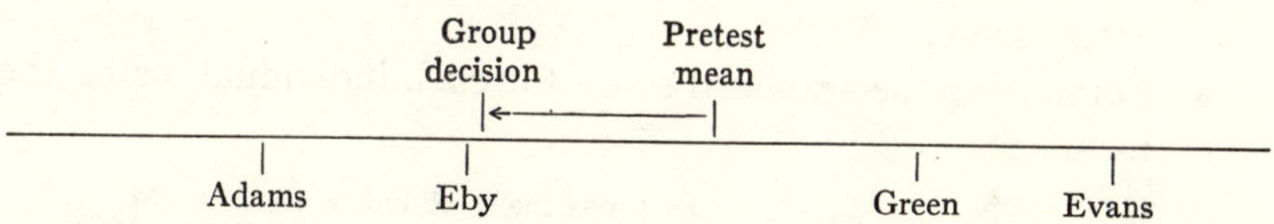

with the weight of group opinion moving, through discussion, in the direction and to the extent shown by the arrow. There is no way of knowing from the data collected whether Adams or Eby was chiefly instrumental in swinging the group in the direction of Alternative A, so each is given the same amount of credit. By an analysis of a sampling of group movements, it was found that in about 17 per cent of the decisions there was a situation comparable to that given above, in which each of two individuals was given the same credit for influencing the group decision. This would indicate that there is a possibility that one person in four is given too much credit and one person in four too little credit about 17 per cent of the time. In other words, there is a possibility that each individual is given some unearned credit about 4.25 per cent of the time; and there is a possibility that each individual was not given enough credit 4.25 per cent of the time. Some of the unearned credit received in some problems by an individual would be counterbalanced by deserved but unreceived credit in other problems. While the writer believes the net distortion for each individual, caused by situations such as that presented above, would be negligible, there is no way of determining its exact amount.

The raw score of an individual's immediate group influence in any particular discussion was found by summating her scores on all the problems discussed by her group.

CONVERTING RAW SCORES INTO STANDARD SCORES. In order to make it possible to add and compare the scores in one series with the scores in another series, it was found advisable to convert the raw scores in each group into standard scores. This was done by working through the following steps for each group:

1. Finding the mean of the scores of each group.
2. Computing the sigma of the scores of each group.
3. Finding the deviation of each individual score from the group score.
4. Computing the standard score for each individual, using the formula:

$$\text{Standard Score} = 50 + \frac{10 \text{ times the x of ind. score from M}}{\text{the sigma of the distribution}}$$

Use of the standard scores instead of raw scores largely eliminated the difficulty caused by the fact that all groups did not finish the discussion of all problems.

ADJUSTING STANDARD SCORES. If a particular student were in a group which was quite strong in so far as total influencing power goes, she would obviously be at a disadvantage. Conversely, if she were in a weaker group, she would get a higher score than she really merited. In order to take care of this differential in the influence potency of various groups, a small adjustment was made in the influence scores for each group in the following manner:

1. The average of the four standard scores for each individual (from the four series discussed) was computed. These may be termed raw standard influence scores.
2. The average influence potency of each group was computed by averaging the raw standard influence scores of the four individuals in the group. This gives a "standard power score" for the group.
3. If the standard power score for the group was above 50, the average of all standard scores, it was concluded that the members of the group had above average influence ability and that a bonus should be given to each individual for competing in a superior group. The amount of the bonus for each individual in that particular discussion was represented by the difference between the group power score and 50. This amount was added to the raw standard score of each individual *for that series.*
4. If the group power score was below 50, then the conclusion was that the group was below average in influencing ability in discussion. So the difference between the group power score and 50 was then subtracted from the standard score which had first been computed for each individual as a measure of her influencing ability in that discussion.
5. The four adjusted standard scores for each individual were then added together and divided by four to give an adjusted final score which represented the individual's average immediate influence ability. This score was used in computing the relationships which will be discussed later.

MEASURING PERSISTING INDIVIDUAL INFLUENCE. As has been noted before, the experiment was designed to measure two types of influence: (1) influence in determining the group decision, or immediate group influence, and (2) influence of one individual on each of the other students in the groups in which the individual worked, sometimes called persisting individual influence. The method used in computing scores for the first type of influence has just been described. We shall now turn our attention to a description of the technique used in evaluating the latter type of influence.

In order to measure the effect of the discussion upon each individual, a retest of each series* of problems was given to each student about five days after the discussion took place, which was about ten days after the pretest had been given. The administration of the latter test was the same as that of the pretest (see page 24). Retest positions were recorded in the same way as the pretest positions.

Thus, using for illustrative purposes the pretest responses for a particular problem of a group that was previously described, and adding the retest positions in a separate column, we have the following picture:

	Pretest position	Retest position
Spragg	15	18
Knott	12	25
Elson	30	26
Jenson	47	37

Since we are interested in estimating the amount of influence each individual has over every other individual, it is necessary to consider the pretest position of each person and see whether each of the other three moved toward or away from that pretest position, the one she would presumably argue for, of the individual we are evaluating at the moment. In this particular case we note with regard to the pretest position of *Spragg* that:

Knott moved away from it 7 more units than she moved toward it, or this = — 7
Elson moved toward her position four points, or this = + 4
Jenson moved toward her position ten points, or this = +10
Thus, the total influence of Spragg on the other three = + 7

Strictly speaking, of course, not all this movement was caused by Spragg, but, in the long run, it would appear that logically this method would give us our best estimate of the influence of Spragg.

Computing *Knott's* score in a similar way we have: Knott's influence on: Spragg = —3; Elson = +4; Jenson = +10. This gives us a total influence score for Knott of +11.

And *Elson's* influence on: Spragg = +3; Knott = +13;

* With the exception of Series I for three classes.

Jenson = +10. This gives a total persisting individual influence score for Elson of +26.

And finally, *Jenson's* influence on: Spragg = +3; Knott = +13; Elson = −4; yielding a total persisting influence score for Jenson on this problem of +12.

On this particular problem it would appear that, relatively, Elson had the most persisting influence, Knott and Jenson about the same, and Spragg the least of all. For each series the raw persisting individual influence score of each discusser was found by adding the persisting individual influence scores of that student from as many problems as were discussed by the group she was in. These raw scores were then converted into raw standard scores by using the same method employed in getting immediate group influence scores. Again also, the method previously described for taking care of differences in influencing power between various groups was used to get the scores finally utilized in later comparisons.

Validity and reliability of the measures of influence. The case for validity of the measures of influence rests largely upon the evidence and logic presented in the description of the methods used in determining influence. Another line of evidence gives a supplementary indication of the validity of the measure of immediate group influence. In two of the five classes which engaged in discussion, the students were asked after each of three of the discussion periods to rank the others with whom they had just discussed, in the order of their probable influence in discussion. This procedure was used in connection with other ranking problems in psychology which had been discussed in class, and care was taken not to lay stress on the factor of influence in the discussions themselves. Each individual used in the correlation computations was ranked by 12 other individuals, which gave a reliability coefficient for the measure of .79. The correlation coefficient of .48 (corrected for attenuation) between these ratings and those obtained by the statistical techniques is indicative of a relationship between the two. A third indication of the validity of the measures involves a subjective evaluation made by the writer. Those who were extremely influential, as indicated by the scales, invariably talked more than the average individual in class discussion and also were more talkative in the small group discussions. Presumably facility in verbal expression would be quite important in discussion and for this reason some importance is attached to this relationship between talkativeness and influence. It is also true that those who practically never volunteered in class, who were comparatively quiet in discussion, invariably received low influence scores.

The statistical reliability of the measures would tend to be reduced by these three factors:

1. The relatively great homogeneity of the individuals with whom the measures were used.
2. The great variety of subject matter employed in the discussions, ranging all the way from the discussion of some specialized art to a discussion of divorce.
3. The time interval occupied in getting the influence measures, almost an entire school semester.

With these influences tending to produce low reliability coefficients for the two types of influence, a coefficient of .54 for immediate group influence was found when the sums of the scores on Series II and III were correlated with the sums of the scores on Series I and IV; and a reliability coefficient of .50 was found for persisting individual influence when scores from the same series of problems were used for comparison. (In the cases where a series score was missing for an individual, because of absence, a slight modification of the split half method was used to get a measure of reliability.)

CHAPTER III

UTILIZATION OF THE McADORY ART TEST

DESCRIPTION OF THE McADORY TEST. This test, which constituted the problems in Series III, consists of 72 pages or plates.* On each plate there are four pictures showing variations of the same subject. The person taking the test is asked to rank the four pictures as he judges them in order of merit from best to worst, recording his response on a record blank provided for the purpose. The complete directions which are placed on top of the record blank are given in the following paragraph:

> DIRECTIONS: Each sheet in these packages shows four pictures called A, B, C, and D. The sheets are named Plate 1, Plate 2, Plate 3, and so on. When you get the package, look at the four pictures on the sheet which says Plate 1 at the top. Pick out the picture you like best. On this record sheet, where it says Plate 1, write on the top line the letter (A or B or C or D) of the picture you like best. On the second line write the letter of the picture you like NEXT BEST. On the third line write the letter of the picture you like THIRD BEST. On the fourth line write the letter of the picture you think is worst. Do the same for Plate 2. Then do Plate 3, and so on. Be sure to put a letter (A or B or C or D) in every space, one for each picture on each plate. [26]

The scoring key for the test has been derived from the rankings as given by experts—critics and teachers—and in scoring an individual's paper a score of one is given for each picture correctly placed according to the key. Norms for each sex at each age to maturity have been established. From figures given in the test manual, [29] it would appear that the reliability of the test is somewhere between .70 and .85, depending somewhat on the age of the individuals tested. The validity of the test will be commented upon later when results achieved through discussion are analyzed.

* Only 68 plates were actually scored since experts could not agree on the order of pictures in each of the other 4 plates.

Purposes in using the art test. The primary purpose in using the test was to obtain two types of influence scores: immediate group influence scores, and persisting individual influence scores. In addition to getting these, the test was also used to determine whether the judgments of students, made as a result of discussion, became more or less like those of the experts who made the test key than had been the previous judgments of the same students.

Procedure. The procedure used in administering the art test was very similar to that used with the other series of problems. (See page 24.) The test was first given to the students individually, using the standard directions and the response sheet which had been made for the test. This took approximately 45 minutes, with some students utilizing the full fifty minutes of the period. Then, five days later discussion of parts of the test in small groups took place, followed by an individual retest five days after the discussion. The students were not aware at the time of the pretest that any discussion was to take place; nor at the time of the discussion were they aware that any retest would be given.

Composition of the groups for discussion. After the pretests were given to each student individually, all pretests were scored. The experts' key was used and a score of one was given for each picture correctly placed according to that key. The students in each class were then ranked according to the size of their scores on the pretest. The individuals were next arranged in groups of four, with each group having approximately the same range of scores. This was accomplished by taking the ranked individuals, in a class of 36 for instance, and placing those numbered 1, 10, 19, and 28 in the first group, those numbered 2, 11, 20, and 29 in the second group, and so on till the nine groups for the class were made up. Thus, each group had in it one individual with a comparatively low score, two with relatively mediocre scores, and one with a comparatively high score.

Questions discussed. In order to find whether "improvement," if any, in score was limited to those questions discussed, it was decided to have each group discuss plates from either the first or the second half of the test, some discussing one half, some the other. Incidentally, "improvement" as used in the preceding sentence and hereafter means greater similarity to the judgments of experts. The test is set up in such a fashion that the second half of the test duplicates, in the general nature of the objects and in the qualities represented, the first half of the test. Thus, utilization of each half of the test as a unit could be conveniently made. In all, fifteen groups discussed parts of the first half of the test, and the same number of groups discussed parts of the second half of the test. At the beginning of the discussion periods the groups were designated and each was instructed to make a group key which represented the best esthetic thinking which the group could muster. Not all plates in the half considered by each group were discussed. The investigator had previously determined those plates upon which the group had agreed in evaluation on the pretest, and the group was instructed to discuss the other plates first, discussing those agreed upon at the end of the period if time remained. It was felt that time could best be spent discussing those plates where there was initial lack of agreement. The number of plates actually discussed by any single group ranged from sixteen to thirty-four.

Obtaining immediate group influence scores. The method used in computing an immediate influence score for each individual on Series III, the McAdory Art Test, will be illustrated by means of the individuals' pretest responses on a particular plate, together with the group responses made after discussion by the group composed of the following four individuals:

	Picture in plate thought:			
	Best	2nd	3rd	Worst
Spragg	A	B	D	C
Knott	A	B	D	C
Elson	A	B	C	D
Jenson	A	B	C	D
Group decision =	A	B	C	D

In the analysis of the plate it will be noted that Spragg and Knott agreed on the pretest as to the proper ranking for the pictures on this plate, but that Elson and Jenson, who agreed with each other, did not agree with Spragg and Knott on the placing of the two lowest pictures. Obviously, discussion of this plate by this group centered on the question: In what order should pictures C and D be ranked? The question we are interested in from the standpoint of influence is: Will Spragg and Knott be successful in getting the group to agree to the order A B D C or will Elson and Jenson carry the point by convincing the other two that A B C D is the proper order? From the group decision it will be noted that, in the consideration of this particular plate, Elson and Jenson won out. The numerical indication of the gain or loss by each individual was computed as follows:

1. The assumption was made that the distance between each picture in the ranking is one unit, which, while not strictly true in every case, was used by the maker of the test in constructing the scoring key, and was found necessary for purposes of this study.
2. With this assumption it was possible to compute from the pretest scores the mean position of each picture on the plate. Mean positions were found to be as follows:

Picture	Position	Picture	Position
A	1	C	3.5
B	2	D	3.5

3. The total deviation of the pretest judgments from the pretest means are then computed for each individual. For instance, Spragg's pretest judgments deviate from the pretest means as follows: From picture A = O; from B = O; from C = .5; and from D = .5. Spragg's total deviation then equals 1 unit. The total deviations of the others computed in the same fashion is as follows: Knott, 1 unit; Elson, 1 unit; Jenson, 1 unit.
4. The total deviation of the pretest judgments from the *group decision* is next computed for each individual. In this problem the results are:

Spragg	A = 0;	B = 0;	C = 1;	D = 1.	Total = 2 units
Knott	0;	0;	1;	1.	Total = 2 units
Elson	0;	0;	0;	0.	Total = 0
Jenson	0;	0;	0;	0.	Total = 0

5. The total deviation for each individual computed in step 4 is subtracted from the total deviation for each individual computed in step 3, and the resulting numbers constitute the immediate influence scores for the four individuals. The immediate influence scores on this plate for the four individuals are: Spragg, —1; Knott, —1; Elson, 1; and Jenson, 1.
6. The final influence score on Series III for each individual is the sum of her immediate group influence scores in the problems discussed by her group, which scores are converted into standard scores in order largely to eliminate the effect of the variation in the number of problems discussed by various groups.

Computation of persisting individual influence scores. The method utilized in getting a score representing persisting individual influence for each student who discussed Series III, the McAdory Art Test, is quite similar to that used with the other series. Since there are more comparisons necessary with Series III than there were with the other series, a comparatively simple example will be used for illustrative purposes. The following table represents the pretest and the retest judgments of the Spragg, Knott, Elson, Jenson group which has been used in describing the method of obtaining immediate group influence scores.

	Pictures in the plate thought:			
	Best	2nd	3rd	Worst
Spragg—pretest	A	B	D	C
retest	A	B	C	D
Knott—pretest	A	B	D	C
retest	A	B	D	C
Elson—pretest	A	B	C	D
retest	A	B	C	D
Jenson—pretest	A	B	C	D
retest	A	B	C	D

In the analysis it will be noted that on this particular plate only Spragg changed from the pretest to the retest; it will also be seen that the change was to the pretest position held by Elson and Jenson. Thus, it would appear that Elson and Jenson should get a positive score on this problem, representing the shift in their direction; also, that Knott should get a negative score, since she lost out in so far as Spragg was concerned; and Spragg should get a score of 0 since none of the other three shifted either toward or away from her pretest position. The persisting influence scores on this problem would be as follows:

Spragg on Knott, 0; on Elson, 0; on Jenson, 0; for a total of 0.

Knott on Spragg, —2 (representing loss of one unit each on pictures D and C since on the pretest the two agreed, but on the retest Spragg moved away from Knott one place for each picture); on Elson, 0; on Jenson, 0; for a total of —2.

Elson on Spragg, 2; on Knott, 0; on Jenson, 0; total, 2.

Jenson on Spragg, 2; on Knott, 0; on Elson, 0; total, 2.

Thus, the scores for influence on individuals for this problem would be for Spragg, 0; for Knott, —2, for Elson, 2; and for Jenson, 2. Of course, from this single problem we do not know whether it was Elson or Jenson who influenced Spragg to change her position on the retest. However, our best assumption is that both were influential, and to the same extent. Theoretically, and apparently in practice, a large enough number of problems will indicate which individual had the most influence.

Total scores for persisting influence in Series III were determined by adding all of each individual's scores on the problems discussed and then converting the scores in each group into standard scores.

Discussion results and experts' judgments. As has been indicated before, the McAdory Art Test was used to get immediate group influence scores and persisting individual influence scores, and was also used to determine whether as a result of discussion the judgments of students tend to become more or less like those of experts (experts here meaning the particular group which helped to validate the test). This was done by comparing the retest score of each student with her pretest score (score being determined by amount of agreement of the student's judgments with those of the experts' key) on both the half of the test discussed and the half not discussed. The exact procedure used will be described in the chapter which gives discussion effects.

USE OF A CONTROL GROUP. Since it might be argued that any gain or loss shown in the retest over the pretest might be due to familiarity with the test, the control group of 24 was given the pretest and the retest in the same fashion as the experimental group, but without the discussion which members of the experimental group participated in. Thus, if there were any tendency for scores from pretest to retest to show a gain which might be attributed to familiarity with the test, or, on the other hand, a loss which might be attributed to boredom with the test, this tendency should be shown by an inspection of the control group's scores on the retest as compared with those they made on the pretest. As a control group the experimenter used students with backgrounds and interests similar to those of the students which were used in the experimental group. The results obtained from this control group were supplemented by results obtained by Siceloff and Woodyard [29] in establishing some other facts concerning the test.

CHAPTER IV

TRAITS AND CHARACTERISTICS STUDIED

THE TRAITS AND CHARACTERISTICS selected for study in order that they might be related to discussion influence of various types may be divided into three categories: (1) those which measure characteristics related to mental ability; (2) those purporting to measure certain personality characteristics; and (3) a miscellaneous group of traits or background characteristics which have no particular relationship to each other. The evaluations which fall under each of these three categories will next be discussed.

I. EVALUATIONS RELATED TO MENTAL ABILITY

1. *The Otis Self-Administering Test of Mental Ability.*[28] Form A of the Higher Examination of this test was used for 126 students while Form B was used for 33 students. The time limit was 20 minutes, as suggested for college students by Otis in his Manual of Directions. This test is designed to measure mental ability which Otis refers to as "that innate mental quality of an individual which increases with age."

2. *The McCall Multi-Mental Test.* [27] This rather novel test, which was constructed to measure the types of intelligence measured by the Stanford Revision of the Binet-Simon Scale and the National Intelligence Test, may be used, according to the directions accompanying the test, for grades three through college. This test was taken by 135 of our experimental students. In the test there are 100 items, each of which consists of five words. The student is instructed to cross out the word which does not belong with the other four. In the scoring, one answer is, for any particular item, considered best; but some of the other possible responses are given a certain amount of credit.

3. *Scholastic Aptitude Tests.* [24] These tests, which are given

annually to approximately 9,000 applicants for admission to college, have been constructed in an attempt to aid colleges in determining the probable success of students in college. In the years 1930 to 1935 inclusive, the scholastic aptitude test was divided into two main subdivisions, a verbal section and a mathematical one. In 1936 the functions of the mathematical section were taken over by the Mathematics Attainment Test, and the verbal section was extended to include five subtests. The scores on the verbal and the mathematical sections are separately compared with influence scores in the use here made of the test results obtained for 118 students.

4. *Semester Marks.* Since class marks give some indication of mental ability, the average semester mark status of each student was computed for the semester in which the discussions she participated in were conducted. In the college in which the experiment was conducted letter grades are given, so that, in order to get a numerical average grade for each student, a subject grade of A was assumed to be worth 4 points, a grade of B worth 3, C worth 2, and D worth 1. These points were totaled for each student and divided by the number of units of work she was taking. This process gives a numerical indication of the scholastic standing of each student during the semester in which the discussions she participated in were conducted.

II. PERSONALITY TRAITS AND CHARACTERISTICS

Various personality tests and inventories which are designed to measure certain personality characteristics have been selected for use in this study. One of the most interesting parts of the experiment was the attempt to determine the extent of the relationship between influence of the types measured and results on certain of these inventories, particularly those which purport to measure dominance in face-to-face situations. In this section, however, we are concerned with giving a picture of the purposes for which these tests were constructed.

The Personality Inventory [22], according to its author, Robert G. Bernreuter, is designed to measure several different aspects of personality at one time. Bernreuter states: "The immediate effect of this is a very considerable saving both in cost and in time required for administration. The facts that the nature of the traits being measured is not readily detectable and that the scales possess high reliability, which permits their being used to compare one individual with another, are further distinct advantages." [23] Six scales have been prepared, four by Bernreuter which are designated by the symbols B1-N, B2-S, B3-I, and B4-D; and two by Flanagan [23] which are designated by the symbols F1-C and F2-S. In Bernreuter's manual these scales are described briefly as follows:

B1-N. A measure of neurotic tendency. Persons scoring high on this scale tend to be emotionally unstable. Those scoring above the 98 percentile would probably benefit from psychiatric or medical advice. Those scoring low tend to be very well balanced emotionally.

B2-S. A measure of self-sufficiency. Persons scoring high on this scale prefer to be alone, rarely ask for sympathy or encouragement, and tend to ignore the advice of others. Those scoring low dislike solitude and often seek advice and encouragement.

B3-I. A measure of introversion-extroversion. Persons scoring high on this scale tend to be introverted; that is, they are imaginative and tend to live within themselves. Scores above the 98 percentile bear the same significance as do similar scores on the B1-N scale. Those scoring low are extroverted; that is they rarely worry, seldom suffer emotional upsets, and rarely substitute daydreaming for action.

B4-D. A measure of dominance. (The items in this test which measure Dominance-Submission are based upon the Ascendance-submission Reaction study by Gordon W. and Floyd H. Allport.) Persons scoring high on this scale tend to dominate others in face-to-face situations. Those scoring low tend to be submissive.

F1-C. A measure of confidence in oneself. Persons scoring high on this scale tend to be hamperingly self-conscious and to have feelings of inferiority; those scoring above the 98 percentile would probably benefit from psychiatric or medical advice. Those scoring low tend to be wholesomely self-confident and to be very well adjusted to their environment.

F2-S. A measure of sociability. Persons scoring high on this scale tend to be non-social, solitary, or independent. Those scoring low tend to be sociable and gregarious.

Maslow "Social Personality Inventory for Women." [25] This test, designed by A. Maslow, represents an attempt to measure the amount of "dominance feeling." Thirty-five items are used in the complete test. These items are distributed among the following factors:

		Number of items
A.	Attitude toward Men	9
B.	Attitude toward Women	4
C.	Self-Evaluation and Behavior	8
D.	Interests and Tastes	8
E.	Conventionality and Miscellaneous	6

Self-Rating Scale.[25] (Accompanying the Maslow "Social Personality Inventory for Women.") By means of six scales this self-rating device gives the student an opportunity to rank herself on the six characteristics related to dominance. By means of the self-ratings on these six scales, a summary self-rating score is acquired which is supposed to represent self-evaluation of dominance feeling.

Classmates' Ratings. After each discussion every member of each discussion group was asked to rank on paper the other three members of the group in which she had been discussing in the order of their desirability as personal friends. Thus, for each student who engaged in each of the four discussions there were obtained twelve rankings by other members of the groups in which she had discussed. This gave a fairly reliable estimate as to the "friendliness" (so called for lack of a better term) of each individual. The coefficient of reliability determined by using an adaptation of the split half method was .84.

Radical Tendencies.[30] By means of the Hartmann Social Attitudes Inventory there was obtained a score indicating the extent of radical tendencies for each individual.

Influenceability. An evaluation of the amount an individual could be influenced or was influenced by discussion was computed by using the amounts of shift on all questions from pretest to retest, a large part of which shift might reasonably be attributed to the intervening discussion. A fuller description of the exact methods used in getting a score representing influenceability is to be found in Chapter VI.

III. MISCELLANEOUS INFORMATION

In addition to getting the above described data, the following information was also obtained for practically all students in the experimental group:

1. Age
2. Religious background
3. Year in school
4. Type of elementary school attended
5. Number of children in family
6. Position in family

In the next chapter the results of comparing influences with the above named traits and background characteristics will be discussed.

CHAPTER V

TRAITS AND BACKGROUND CHARACTERISTICS COMPARED WITH THE TWO TYPES OF INFLUENCE SCORES

IN CONSIDERING THE RELATION of particular traits or characteristics to influence in discussion, a distinction should be made between two types of relation which might exist. To illustrate the first type let us assume that a particular individual has an excessively confident, self-assured manner. It might be found that in some situations this would produce for him a very influential part in determining the position to be recommended by the group regarding a particular problem; on the other hand, it might be found that under other circumstances this same characteristic which had produced for him an influential place in formulating a group decision would antagonize the other discussers so that his influence would be extremely limited. Obviously, the self-confident, self-assured manner was a factor of great importance in determining this individual's influence in discussion. If the same thing were found for other individuals and it could be substantiated that such a characteristic was a disadvantage about as often as an advantage to individuals possessing it, we would say that the characteristic is certainly important in determining influence in discussion, but that knowledge of its possession by an individual will not help us in predicting his probable influence in discussion (unless we are able to determine adequately the kind of group and the type of situation in which it tends to work in a particular way). In this study an investigation was not made of this first type of relationship.

Rather it was attempted in this study to find out whether the possession of a large amount of a particular characteristic tended to be *generally or consistently* associated in either a positive or a negative fashion with much influence in the types of situations encountered in the experimental discussions. In other words, no attempt was made to determine all factors related to influence in discussion in the sense in which relationship was illustrated in the preceding paragraph, but an attempt was made to determine whether certain factors generally tended to be related in a consistent fashion to influence in discussion. Thus, it should be made quite clear that simply because little or no correlation is found between dominance, for instance, as measured by pencil and paper tests, and influence in discussion, it does not mean that the possession of the trait measured does not affect influence in discussion, but rather it means that the effect, if any, was not either to raise or to lower influence scores consistently.

In this connection it should be pointed out that influence itself is not to be thought of as a unitary trait of which an individual can possess a certain amount. It is rather to be considered as something which is to a very large extent dependent upon the type of situation in which the individual finds himself. In one situation influence in discussion may depend largely upon one thing and in another situation largely upon something else. Even in the partially controlled classroom situation under which this experiment was conducted, it will be noted that the reliability coefficients of the measures of influence are comparatively low although each individual's score was based on a sampling of the effect of more than three hours of his discussion behavior. The multitudinous factors whose complex interaction may be thought of as determining the individual's influence in a particular discussion when recombined in a new situation may produce a quite different result. This obviously makes it extremely difficult to get what may be thought of as an adequate sampling of the influence an individual will have in discussion. Nevertheless, measures were obtained which showed considerable consistency. In this chapter we shall show the correlation of these measures of influence with certain individual characteristics.

In studying the relationship between influence and other factors three methods have been employed. These three methods are (*a*) the Pearson product-moment correlation method; (*b*) the significance of the difference-between-means technique; and (*c*) a recently devised method which makes it possible to compare on certain basic traits two groups having different distributions and different means. The theory back of this latter method is described by P. O. Johnson and J. Neyman in *Statistical Research Memoirs,* I (June, 1936), 57-93, under the title, "Tests of Certain Linear Hypotheses and Their Application to Some Educational Problems."

Relation of scores on certain tests of mental ability to influence scores of two types

The Otis Self-Administering Test of Mental Ability. When scores on the Otis Test were correlated with scores representing amount of immediate group influence, the resulting coefficient of correlation was $.02 \pm .05$ (this and all subsequent correlation coefficients have been corrected for attenuation); and when Otis scores were correlated with persisting individual influence, the correlation coefficient was $-.13 \pm .05$. No reliable relationship was demonstrated. (As a matter of interest the Otis Test scores were correlated with the average semester mark and a coefficient of only .08 resulted. Apparently marks and scores on the test are measures of two things which are almost entirely independent.)

Multi-Mental Test. When Multi-Mental scores of the 135 students who took the test were correlated with the same students' immediate group influence scores, there resulted a coefficient of correlation of $.11 \pm .06$. When persisting individual influence scores were correlated with Multi-Mental scores, the resulting correlation coefficient was $.07 \pm .06$. No significant relationship is indicated in either case.

Scholastic Aptitude Test of Verbal Ability. This test, it will be noted, is a rather comprehensive test given to applicants for admission to colleges, and has been validated largely on scholastic standing as indicated by teachers' marks. When scores of 118 experimental students who had taken this test were correlated with immediate group influence scores, there resulted a coefficient of correlation of $.45 \pm .06$. This coefficient, which is seven and one-half times its probable error, indicates a substantial relationship between the type of mental ability measured by the Scholastic Aptitude Test of Verbal Ability and the type of ability which tended to characterize individuals who are very successful in influencing others to subscribe, temporarily at least, to their points of view in discussion. This test, when correlated with persisting individual influence, yields a coefficient of correlation of .29± .06. This latter coefficient, while not as high as the preceding one, is still almost five times its probable error, which indicates a significant relationship existing between the Aptitude Test of Verbal Ability and the measure of ability to influence others in a discussion group so that the influence persists. As a matter of interest, this Scholastic Aptitude Test was correlated with school marks for a semester and there resulted a coefficient of correlation of .33±.06, which it will be noted is lower than that between immediate group influence and marks.

Scholastic Aptitude Test of Mathematical Ability. When the results from 102 students for this test of mathematical ability were correlated with immediate influence, the resulting coefficient of correlation was .22±.066. This would indicate that the chances are about 99 out of 100 that a positive relationship really exists between mathematical ability as measured by this test and immediate group influence. The results from the test, when correlated with persisting individual influence scores, give a correlation coefficient of .36±.066. This coefficient, which is about five and one-half times its probable error, would make it appear quite certain that there is a relationship between mathematical ability as measured by this Aptitude Test and influence of the "persisting sort." This mathematical section of the Scholastic Aptitude Test was also correlated with school marks for the semester in which the discussion took place. The coefficient of correlation between these two measures was .17±.07, which is considerably lower than that between mathematical ability and persisting individual influence, and approximately the same as that between mathematical ability and immediate group influence.

Semester Marks. When the semester marks for the 161 students in the experimental group were correlated with immediate group influence, a correlation coefficient of .38±.05 resulted. This coefficient, which is more than seven times its probable error, would indicate that students who are best able to sway their classmates in discussion are, on the average, also best able to sway their teachers and favorably impress them through classroom work, examination papers, office interviews, etc., so that they get the best marks. In the college in which the experiment was conducted marks are largely based on written work rather than on oral work, so it is interesting to see the relationship between influencing classmates through oral discussion exclusively, and influencing teachers largely through written work. When class marks were correlated with persisting individual influence, the resulting coefficient of correlation was $.19 \pm .05$. This coefficient with its probable error indicates that the chances are about 99 in 100 that a positive relationship exists between marks and influence of a persisting sort.

Relation of scores on certain personality indices to scores representing influence

Bernreuter and Flanagan Personality Inventory Scales. In considering the results obtained in correlating influence scores with various personality measures, it will be well to bear in mind that both the influence measures and the personality measures are admittedly somewhat crude. Some psychologists tend to feel that the manifestation of such traits as dominance and self-sufficiency is probably too much a part of a specific situation to be measured in a significant way by general tests. If this is true, it will probably be advisable to indicate to just what kind of situation the test is expected to apply.

The results of correlating immediate group influence with neurotic tendency, self-sufficiency, introversion-extroversion, dominance-submission (all from the Bernreuter scales) and self-consciousness and non-sociability (both from the Flanagan scales) are summarized on Table I, as are also the results from correlating the above named six scales with persisting individual influence scores. All coefficients are derived from 161 cases and have been corrected for attenuation.

It will be noted that only one of the twelve coefficients given on

Table I approaches statistical reliability. This one of —.19 between self-consciousness and immediate group influence indicates the chances are about 99 out of 100 that self-consciousness is inversely related to immediate group influence.

TABLE I

CERTAIN PERSONALITY MEASURES CORRELATED WITH (1) IMMEDIATE INFLUENCE SCORES, AND (2) PERSISTING INDIVIDUAL INFLUENCE SCORES*

	With immediate group influence	With persisting individual influence
Bernreuter Scale of Neurotic Tendency	—.04	.05
Bernreuter Scale of Self-Sufficiency	.08	—.03
Bernreuter Scale of Introversion	—.09	—.01
Bernreuter Scale of Dominance	—.02	—.06
Flanagan Scale of Self-Consciousness	—.19	.02
Flanagan Scale of Non-Sociability	.01	—.02
Maslow Test of Dominance Feeling	.13	.02
Self-rating of Dominance Feeling	—.02	—.06
Radicalism	.18	.10
Likability as rated by classmates	.18	.23
Influenceability or amount influenced by discussion	—.41	—.34

* Number of subjects in all cases was 161. Probable error of all coefficients = ± .05.

The coefficients of correlation based on B4-D and the two types of influence are particularly interesting. B4-D is called a measure of dominance and persons scoring high on it are supposed to "dominate others in face-to-face situations. Those scoring low tend to be submissive." In determining those who would be most influential in discussion, it would appear from this description that the B4-D is made to order. Thus, one would be justified in expecting a fairly high correlation between scores on this scale and influence scores, particularly immediate group influence scores, for here surely we have a specific face-to-face situation and a very real opportunity for dominance to be shown. However, such expectations definitely are not realized. The coefficient of correlation between B4-D and immediate group influence is —.02; and that between B4-D and persisting individual influence is —.06. An explanation of these small negative coefficients of correlation appearing where fairly high positive coefficients might be expected may rest in some weaknesses of the measures of influence, whose reliability and validity have already been discussed; on the other hand, this result may be due to the method of validation of B4-D. The items of this scale are based upon the ascendence-submission reaction study by Gordon W. and Floyd H. Allport [21]. The Allports' scale in turn was not validated on any direct criterion of dominance or ascendence but only upon the average of five student ratings (four by other students and one by the student himself) of ascendence on the individuals whose scores were used in validating the Ascendence Scale. Thus, we have the B4-D scale built upon the Allport Scale, which in the first place was based only upon ratings whose validity is somewhat doubtful. If the original ratings on ascendence have little validity (which may or may not be the case), then the Allport Scale itself would have little validity, and one would expect the B4-D to have even less. At any rate, discussion dominance of the type we have measured seems to be totally unrelated in a consistent fashion to the type of dominance being measured by this test of general dominance.

Maslow Social Personality Inventory for Women. This is another of the general dominance tests. It is designed to indicate amount of "dominance feeling." Correlation between immediate group influence and results on this test yields a coefficient of .13; correlation between persisting individual influence and the Maslow Test gives a coefficient of .02. The latter coefficient, of course, establishes no relationship, while the former would indicate that the chances are 96 out of 100 that there is real relationship between immediate group influence scores and Maslow Dominance scores.

Self-Rating Scale. This scale, given with the Maslow "Social Personality Inventory for Women," is supposed to represent a self-evaluation of "dominance feeling." High scores indicate a great amount of dominance feeling, while low scores indicate a lack of dominance feeling. Correlation of scores on this sheet with immediate influence scores yield a correlation coefficient of —.02; while scores on this sheet when correlated with persisting influence scores give a coefficient of —.06. Of course, with a probable error of .05, neither of these coefficients demonstrates any relationship between amount of dominance feeling and amount of influencing ability.

Radicalism. It will be remembered that the "Hartmann Social Inventory," which is really a test of radical tendencies, was given to the experimental subjects, largely to see whether the theory which holds that radicals are more energetic and influential than conservatives would be borne out. When immediate influence was correlated with radicalism as measured by this test, a coefficient of correlation of .18±.05 resulted. This would indicate that the chances are about 99 out of 100 that a positive relationship exists between the two variables correlated. The coefficient of correlation between radicalism and persisting individual influence turned out to be .10±.05, which probably indicates a slight relationship, but certainly does not prove that anything but the vagaries of selection were operating. Thus, there is some evidence to support the theory that individuals with radical tendencies are somewhat more influential than individuals with conservative leanings.

Classmates' Ratings. Classmate rating scores indicate the average desirability of each individual as a personal friend as shown by the combined rankings made by those who worked with each student in discussion. Presumably those with most influence, at least influence of a persisting sort, would be those who are best liked. When general desirability scores were correlated with immediate influence scores, a correlation coefficient of .18 resulted, indicating that the chances are about 99 out of 100 that a real relationship exists between the two variables. A correlation coefficient of .23 resulted when general desirability scores were correlated with persisting influence scores. Interpreted, this would mean that the chances are 999 out of 1,000 that a real positive relationship exists between being desired as a personal friend and influence on individuals in discussion.

Influenceability Ratings. It will be remembered that what have been called influenceability ratings represent estimates of the relative amounts individuals were influenced by the discussions in which they participated. A high score indicated that the individual was influenced a great deal, while a low score indicated that the responses of the individual on her retest were influenced by the discussion comparatively little. From a theoretical standpoint one would probably expect that those who were most influential would themselves be least influenced in the discussion. When immediate influence scores were correlated with influenceability ratings, a coefficient of —.41 resulted; while when persisting influence was correlated with influenceability, the coefficient of correlation was —.34. The theoretical expectation is thus rather strongly borne out by investigation and we conclude that those who are most influential, are on the average, least influenced.

The results which have been discussed in the last five paragraphs are summated at the bottom of Table I which appears on page 56.

Miscellaneous data compared with influence

In comparing the two types of influence scores with various background characteristics, two statistical devices have been employed; (1) the significance of the difference between means technique; and (2) a recently devised method by means of which two groups having different distributions and different means may be compared on one trait after partialing out the effect of certain background characteristics which might be expected to influence the difference between the groups in the trait being studied.

Year in School. In two classes of the experimental group there was a sufficiently large number and a sufficiently even distribution between two groups, the one being made up of freshmen and sophomores and the other being composed of juniors and seniors, to justify making a comparison between the two groups. In the first group, composed of first and second year students there were 38 subjects; in the second group there were 27 third and fourth year students. When the mean immediate group influence score of the freshman and sophomore students was subtracted from the mean influence score of the juniors and seniors, there resulted a difference of 1.80 points, which when divided by the sigma of the difference between means yielded a critical ratio of 1.32. Interpreted this would indicate that the chances are 91 in 100 that there exists a real difference in immediate influence favoring the upper-class students.

When persisting individual influence was compared with the two groups in a similar fashion, a critical ratio of 3.52 resulted. This indicates that the factor of chance alone could not have produced the average superiority of juniors and seniors over freshmen and sophomores in influencing ability of a persisting sort.

Single Child vs. Children with Siblings. When the mean immediate group influence score of 46 single children was compared with the mean score of 115 subjects who have one or more siblings, the resulting difference of 2.66 points favors the single child. This difference when divided by the sigma of the difference between means yielded a critical ratio of 2.51. Interpreted, this ratio would indicate that the chances are 994 out of 1,000 that there is a real superiority of single children over children with siblings in ability to influence the formulation of the group decision.

The difference between the persisting individual influence means of the two groups when divided by the sigma of the difference produced a critical ratio of 2.26. A ratio of this size indicates here that the chances are 988 out of 1000 that single children in populations of this type have more influence of a persisting sort than do children with siblings. These two differences, which closely approach statistical reliability, might be explained by the fact that single children have had more opportunity in the family circle to express themselves. The differences might also be partially due to the single child's having her way more often in the home, although this would probably not affect the persisting individual influence score.

First Child vs. Other Sibs, Singletons excluded. The first child group was made up of 52 students who were first children and had siblings, while the "other sib" group was made up of 60 students who had an older sibling or older siblings in their families. When the mean score of the first child group is compared in immediate group influence with the mean score of the other sib group, the difference divided by the sigma of the difference between means gives a critical ratio of .70, which would indicate that the chances are only about 76 in 100 that there is a real difference favoring children who are not the oldest in their families in the populations of which these groups are samples.

Using persisting individual influence scores, there results a difference in the other direction, favoring the oldest. The critical ratio of 1.40 in this case would indicate that the chances are 92 out of 100 that there is a real difference in persisting individual influenceability favoring the oldest child over the child who has an older sibling. Neither difference is statistically reliable as the chances indicate.

Public vs. Private Elementary School Products. In the private school group were placed all those who reported any private school experience (with students from parochial schools, four in number, excluded) in their elementary school training; in the public school group were included all those who had spent their entire time on the elementary school level in the public schools. When the mean immediate group influence score of students, all of whose elementary experience was in the public schools, was compared with the mean score of students who had had private elementary school experience, there resulted a slight difference favoring the private school group. The difference divided by the sigma of the difference between means yielded a critical ratio of 1.74, which would indicate the chances are about 96 out of 100 that there is a real difference favoring the private school population. (All four students trained in parochial elementary schools had below average scores. Four scores are not very significant, of course, but are mentioned as a matter of interest.)

Using persisting influence scores of these two groups, we find a slight difference between means which again favors the private school group. The critical ratio of .70 indicates that the chances are only 76 in 100 that there is a real difference favoring the private school group. Neither ratio given above indicates a statistically reliable difference, with the immediate group influence, one, however, approaching reliability. (The four parochially trained students again had scores which averaged considerably below average.)

Progressive vs. Traditional Training. On a questionnaire three classes of students were asked to indicate the number of years of their school career that had been spent in a "progressive school." As an aid to the students in doing this, half a dozen criteria of "progressivism" were indicated. Admittedly, the responses are not, could not be, particularly accurate. However, the results are given for what they are worth. The progressive group was made up of those who had had at least four years in a progressive school; the traditional group of those who had had all training in traditional schools. Theoretically, the progressive school products would probably be expected to have the most influence in formulating the group decision because of the greater emphasis in the progressive schools on expression and initiative. The difference between the means of the two groups on immediate group influence slightly favor those who had had, according to their own testimony, at least four years in a progressive school. The difference divided by its standard error yielded a critical ratio of 1.75, which indicates that the chances are 96 out of 100 that there is a real difference in the populations represented favoring the progressive products.

However, with persisting individual influence scores, the difference between means, though very small, does favor traditional training. The difference divided by its standard error gives a critical ratio of only .70, indicating that the chances are about 76 in 100 that there is a real difference between the two populations of which these groups are samples.

Religious Background. In view of the fact that different religious groups in this country give different emphasis to such things as initiative, obedience, respect for elders, etc., it is interesting to see the relation of differing religious backgrounds to influence in discussion. First, comparing the representatives of the two largest religious groups in the country, Protestants and Catholics, we find the mean immediate group influence score of the Protestant group to be somewhat higher than the mean score of the Catholic group. The difference when divided by its standard error produces a critical ratio of .98, indicating that the chances are 81 in 100 that there is a real difference in the populations from which our groups are samples. However, the mean persisting influence score of Catholics is almost exactly the same as that of Protestants, indicating that the chances are about 50-50 on the average for either group to excel in this latter type of influence. Of course, neither difference is statistically reliable.

When the mean immediate group influence score of Protestant students is compared with that of Jewish students, the resulting difference favors the Jewish group. This difference when divided by its sigma yields a critical ratio of 2.55, indicating that the chances are 995 out of 1,000 that there is a real difference favoring Jewesses in the populations from which these samples are drawn. The difference between means of persisting individual influence scores again favors the Jewish group, although the difference is not reliable. The critical ratio in this case is 1.14, which indicates the chances are about 87 out of 100 that there is a real difference favoring Jewish students in the represented populations. While the latter difference is not statistically reliable, the former one approximates statistical reliability.

Comparison of the mean immediate group influence score of Jewish students with that of Catholic students produced a comparatively large difference favoring the former group. The critical ratio equals 3.00, indicating that the chances are 999 in 1,000 that there is superiority, on the average, of Jewish students over Catholic students. In addition to the fact that Catholic families place great emphasis upon obedience as opposed to aggressiveness in children, it should also be noted that the Jews constitute a minority group which has frequently been considered characteristically aggressive. When persisting individual influence means of the two groups were compared, the difference again favored the Jewesses, although it was not found to be as large with this type of influence. The critical ratio found by dividing the difference between means by the standard error of the difference was 1.47.

In addition to using the difference between means technique in comparing these two groups with regard to immediate influence, the investigator also used a recently devised method which allows the utilization of the full distributions of the two groups without rejecting cases to secure matching, and yet effect a comparison of the two groups undistorted by discrepancies in certain related background traits which are believed to condition the size of scores on the trait on which we are comparing the two groups. This is achieved by a somewhat laborious process of subdividing the groups into small subclasses which are relatively homogeneous with respect to the background traits, so that comparison is feasible within each of these subclasses. When properly weighted, the sum of the intra-class differences may be compared with the totality of intra-class variability. In this particular case the factors whose effect was largely eliminated by the process were grades in school and year in school. In other words we were attempting to find out whether with the two groups, Jewish and Catholic, roughly equated for year in school and grades in school, there would still be a reliable difference favoring the Jewish group. Without describing the process, which is rather involved and whose source has already been mentioned,* we will reiterate the conclusion which is that even with the groups equated for year in school and for school marks there still was a significant difference favoring the Jewish group. The evidence just given may be very briefly summarized as follows:

SUMMARY

1. In general, those with the greatest amount of immediate group influence also tended to excel others in the amount of persisting individual influence exerted; and conversely, those with little immediate group influence tended to have little persisting individual influence.

2. Those with most immdiate group influence as well as those with most persisting individual influence had more scholastic ability, were less self-conscious, were more desired as personal friends by classmates, held slightly more radical ideas, and were less influenced by discussion.

3. No statistically reliable relationship was established between either immediate group influence or persisting individual influence on the one hand and either neuroticism, self-sufficiency, introversion, dominance as measured by general pencil and paper tests, or sociability, on the other hand.

* Page 53.

4. Neither group in each of the following pairs was found to excel significantly the other group in either type of influence: public elementary school products vs. private elementary school products, first child vs. siblings with an older brother or sister (singletons excluded), progressive vs. traditional school products.

5. Religious background was found to be a significant factor, with those having a Jewish background exerting most immediate group influence, Protestants in the middle, and those with Catholic training having the least immediate group influence. Jewesses also had the most persisting individual influence with Protestants and Catholics having about the same amount.

6. Juniors and seniors had more influence of both types than did freshmen and sophomores.

7. Single children had more of both type of influence than did students with siblings.

CHAPTER VI

RESULTS CONCERNED WITH INDIVIDUAL SHIFTS FROM PRETEST TO RETEST

WHILE THIS STUDY WAS DESIGNED primarily to determine the type of individual most influential in discussion, it was also possible to treat the data gathered in such a way that a measure of what has been termed the influenceability of each student could be obtained. Influenceability is the term used to designate the amount an individual shifted in his responses from pretest to retest, at least a part of which shift could reasonably be attributed to the discussion engaged in by the student. On the educational, social, economic, and political issues the amount of individual shift on each question was determined by using the scale of 50 units previously explained in describing the measurement of influence. The following illustration will make clear the method employed: If an individual had position *P* on the pretest and position *R* on the retest, as represented on the following scale, then his influenceability score on that question is represented by the distance

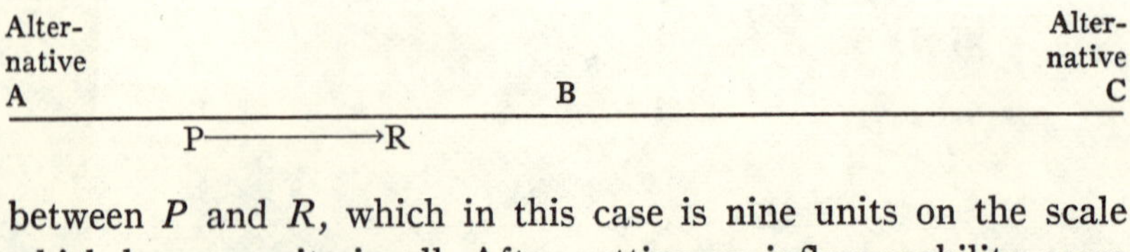

between *P* and *R*, which in this case is nine units on the scale which has 50 units in all. After getting an influenceability score for each individual in a group, which discussed a particular series of problems, the raw scores were converted into standard scores for the members of each group by the method which has been described in connection with the explanation of how immediate group influence scores were computed. (See page 34.)

In the case of the Art Test shifts from pretest to retest were represented by the change in rank given to each picture on the 68 plates considered in the test. For instance, if an individual ranked the pictures on a particular plate in this order on the pretest: D A C B, and this order on the retest: C D A B, then picture D changed its ranked position from first to second, a change of one, picture A from second position to third, again a change of one, picture C from third to first, a change of two, while picture B remained in the same position. Thus, the total number of shifts on this plate was 1 + 1 + 2 + 0, or 4. Standard scores were then computed by groups for the series.

The standard scores for each individual were then averaged. The resulting average gave what will hereafter be termed the influenceability score.

Influenceability scores when correlated with each of the following test scores gave no results which indicated a significant relationship between influenceability and the characteristic measured: Otis Mental Test, Multi-Mental Test, Scholastic Aptitude Test of Verbal Ability, Scholastic Aptitude Test of Mathematical Ability, and Semester Marks. The actual coefficients of correlation are given on Table II.

TABLE II

CORRELATION COEFFICIENTS RESULTING FROM CORRELATING INFLUENCEABILITY WITH CERTAIN OTHER CHARACTERISTICS

Influenceability scores correlated with		No. of cases
Otis Test	.05	159
McCall Multi-Mental Test	—.11	135
Scholastic Aptitude Test of Verbal Ability	.10	118
Scholastic Aptitude Test of Mathematical Ability	—.09	102
Semester Marks	.02	161
Self-sufficiency	—.08	161
Radicalism	—.02	161
Classmates' rating on likability as a personal friend	—.29	161

Nor was any significant relationship established between influenceability and either of the following characteristics: self-sufficiency as measured by the Bernreuter scale, or radicalism as measured by the "Social Attitudes Inquiry." When desirability as a friend as measured by classmates ratings was correlated with influenceability, a coefficient of —.29 resulted. This coefficient, based on 161 cases, indicates a significant negative relationship between influenceability and desirability as a friend. Coefficients of correlation for these comparisons are also given in Table II. It may be that no relationship exists between being influenced by discussion and the characteristics mentioned in this and the preceding paragraph, with the exception of desirability as a friend; or it is possible that the variability in the effect of discussion has been counterbalanced or overshadowed by variation produced by other causes besides discussion.

Using the difference between means technique, the following pairs of groups were compared to see whether either group of a pair had significantly higher influenceability scores: Public elementary school products vs. private school products, progressive vs. traditional school products, students with Protestant religious background vs. students with a Jewish religious background, students with a Protestant religious background vs. students with a Catholic religious background, students with a Jewish religious background vs. students with a Catholic religious background, freshmen and sophomores vs. juniors and seniors, and single children vs. children with siblings.

In only three pairs of cases was there found a difference between the means of the two groups in each pair which approached statistical significance. Catholics and Protestants both were more influenced than Jewesses. The chances that these differences are real in the populations from which these samples are drawn are 99 out of 100 in the case of Catholics compared with Jewesses, and 95 out of 100 when Protestants are compared with Jewesses. Students who had had at least four years in a progressive school made more shifts from pretest to retest, with discussion intervening, than did those with traditional school background. The chances are 96 out of 100 that the found difference in these samples is indicative of a real difference giving higher average influenceability scores to students coming from progressive schools.

In general, it appears that those groups (as well as those individuals) who have more influence in discussion than the average are somewhat less influenced than the average by discussion.

Do extremists change most through discussion? The question next to be discussed is whether those with the most extreme views on the subjects discussed change their views more from pretest to retest than do those with more moderate views. All responses on both pretest and retest to questions in Series V, VI, VII, and VIII were investigated in an endeavor to answer this question. Each problem in each group was treated separately, and the results were later summated for each problem and for each series. On each problem in each group the amount of shift of the person with the most extreme view was recorded as was that of the person in the group who held the most moderate view as indicated by her pretest response. These two contrasting scores for a problem, the most extreme and the least extreme, were separately summated for all groups. No significant difference was found between the amount of shift by individuals with extreme views and those with more moderate views. In two series the results gave slightly higher shift scores to extremists and in the other two series the "middle-of-the-roaders" shifted more from pretest to retest.

This result, coupled with the lack of correlation between radicalism and influenceability, would seem to indicate that individuals with more moderate views are probably influenced as much by discussion as are those with more extreme views.

Individual shift with relation to the opinions of experts. The reader will recall that all students first took the complete Art Test individually; that five days later each student participated in a discussion group which considered a majority of the plates in only one half of the test; and that five days after the discussion each student again took the complete test, the latter performance being called the retest. This sequence of events was not known to the students beforehand. From this setup it was possible not only to compare the total score on the pretest for each student with her score on the retest, but also to compare the amount of gain or loss shown on the half of the test which was discussed with the gain or loss made on the half of the test not discussed (gain here indicating greater agreement with the experts' judgments). The retest, as used in this connection, helps us, then, to answer the following two questions with respect to the population with which we are dealing: (1) Does discussion of esthetic problems with one's peers improve one's judgments with regard to the artistic problems discussed? and (2) Does the discussion of esthetic problems with one's peers improve one's judgments with regard to esthetic problems similar to those which were discussed? The control group is also of great value in this connection, since in this group no discussion took place between pretest and retest.

Results of discussion of artistic subjects upon students' subsequent judgments on problems discussed. The average pretest score of students, based on the experts' key, on the half of the test which was later discussed was 89.73; the average score of these students upon this half of the test taken ten days later, five days after discussion, was 98.28. This indicated an average gain from pretest to retest on the half of the test which was discussed of 8.55.

In order to interpret the significance of this apparent gain, it is necessary to know what effect, if any, mere familiarity with the test, as gained by taking the pretest, would have on the score on the retest. The control group has been used to give an answer to this question. The average pretest score on the complete test of the control group was 180.8, while the average retest score was 181.8. This gives an average gain for the control group of one point on the complete test, or of one-half point on one half of the test.

Thus, if we assume that with our experimental group an average gain of .5 on the half of the test discussed would have been made without any discussion, as the results on the control group would indicate, then the amount of gain due to discussion on the half of the test discussed would be represented by 8.55 minus .5 or a net gain of 8.05 points. Using the formula

$$\sigma_{\text{diff}} = \sigma m^2_p + \sigma m^2_r - 2r p \sigma m_p \sigma m_r,$$

the sigma of the difference between pretest and retest will equal .81. Dividing this sigma of the difference into the observed difference of 8.05 results in a critical ratio of 9.94. Interpreted, this would of course indicate that there is a very significant gain in score, far beyond what might be expected by chance from pretest to retest on the half of the test which was discussed. The question may, of course, be raised as to the validity of the test with the key made by a particular group of judges. However, using the key made to go with the test, we found that discussion led students to make judgments more similar to those of experts than they would have done without the discussion.

Results of discussion upon subsequent judgments involving esthetic problems similar to those discussed. Each of the 118 students who was in a group which discussed problems from either the first or the second half of the test took a complete retest. The average pretest score of those students on the half of the test not discussed was 89.97; the average retest score on the half of the test not discussed was 95.42, which gives a gain from pretest to retest of 5.45. Subtracting the .5 gain which might be expected because of familiarity with the test, there results a net gain of 4.95 points which may be called a transfer gain due to discussion of similar problems in the other half of the test. The sigma of the difference between the pretest and the retest equals .81. Dividing 4.95, the observed difference, by .81, the sigma of the difference, there results a critical ratio of 6.11. Thus, with an observed difference of more than six times the standard error of the difference, there is unequivocal evidence that with this student population, judgments involving certain esthetic problems become, as a result of the discussion of similar esthetic problems, more similar to those of experts than they would have been had not the discussion taken place.

Siceloff and Woodyard report of gains from pretest to retest. Siceloff and Woodyard, in *"Validity and Standardization of the McAdory Art Test,"* [29] report (page 27) the gains made by a group of 59 Hunter College women who took a retest one week after the initial test. In this group, which had instruction in art during the interval, the mean increase was 2.4 points, after the one-week interval. It is of interest to note also that another group consisting of 47 women college students who were given a retest one year after the pretest, with art training during the year, had a mean increase of 4.6 points in retest over pretest. These results do not have any direct bearing on the present experiment but are given as a matter of interest.

Conclusions regarding effect of discussion on judgments. The answer to each of the questions raised at the beginning of this section: (1) Does the discussion of esthetic problems with one's peers improve one's judgments with regard to the esthetic problems discussed? and (2) Does such discussion of particular problems improve one's judgments with regard to other, but similar, esthetic problems? is, for this population at least, definitely yes.

Character of esthetic judgment and amount of shift. While it has been demonstrated that there was a significant average shift toward the opinion of experts from pretest to retest as a result of discussion, there remain the questions: What individuals within the groups made the greatest shifts? and What individuals made the greatest gains? In other words, was the size of score on the Art Test related to either the amount of shift or the amount of gain resulting from discussion? We shall first take up the question of the amount of individual shift.

On the half of the test discussed the average shift of the highest individual in each of the 27 groups of four students was 44.8; the average shift of the next to the highest individual in each group was 51.2; that of next to the lowest was 58.0; and that of the lowest was 68.1. The comparison of these averages is shown graphically by the unbroken line in Table III. It will be seen that, in general, those with the highest scores made the smallest number of shifts on the half of the test, the larger part of which had been discussed.

Not only was this trend noted on the half of the test discussed, but it was almost equally evident on the half not discussed. The average shift scores which were, from highest individual to lowest, 38.44, 49.3, 57.1, and 65.3, are graphically depicted by the middle line in Table III. In other words, an inverse relationship was found to exist between size of score on the pretest and amount of shift made from pretest to retest. The reason for this relationship probably is that those who make the highest scores have more definite opinions, more crystallized ideas, than those who make comparatively low scores. Those with low scores apparently have many opinions which are on the threshold of change.

Size of score of control subjects related to amount of shift. In order to see whether the same relationship between size of score and amount of shift would be apparent even if discussion did not

TABLE III

SHOWING AVERAGE SHIFT BY STUDENTS ON VARIOUS LEVELS

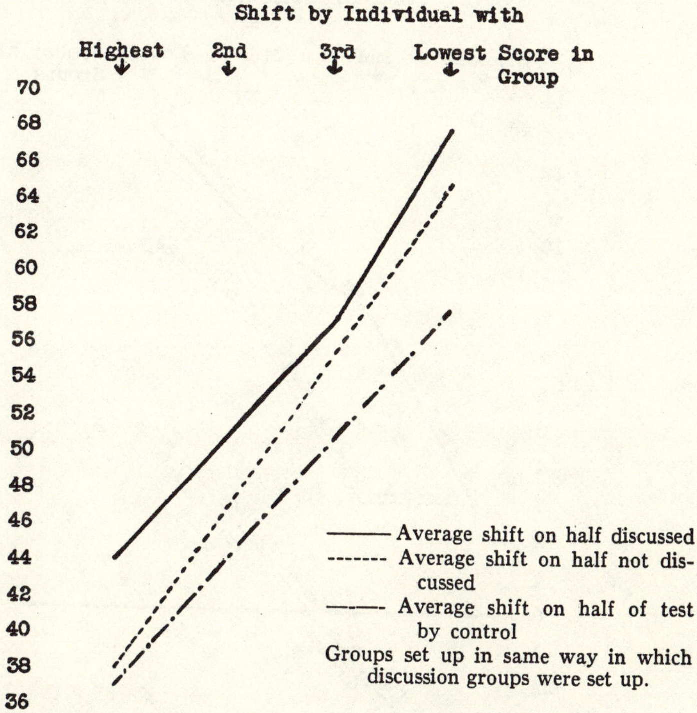

take place, the scores of the control subjects from pretest to retest were analyzed. The 24 control subjects' pretest scores were first arranged in descending order. Then six groups were made up in exactly the same fashion as the experimental groups had been made up except that of course the control subjects did not actually meet. Then the average shift from pretest to retest for the twelve highest in the groups was compared with that of the twelve lowest, with the results pictured in the lowest line in Table III. It will be seen that control subjects with the highest scores on the pretest made the least shifts and that when the line between the two groups is extrapolated at both ends on the table, the inclination is about the same as that of the other two lines on the table.

TABLE IV

Average Gain or Loss from Pretest to Retest on Half of McAdory Art Test by Members of Groups

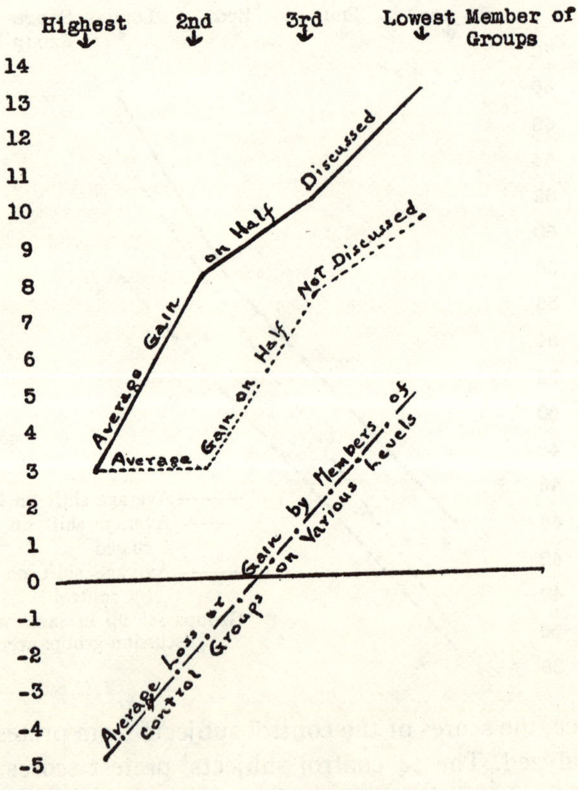

Thus it would appear that the comparatively small amount of shift from pretest to retest by those with high scores in the experimental group is largely independent of discussion.

Relation between size of score and amount of gain. As has been indicated before, the average gain of those students who engaged in discussion was 8.55 points on the half of the test discussed and 5.45 points on the half of the test not discussed. On analyzing the gain of individuals in the same group with differing pretest scores,

TABLE V

AVERAGE GAIN OR LOSS FROM PRETEST TO RETEST OF EACH OF FOUR LEVELS ON MCADORY ART TEST

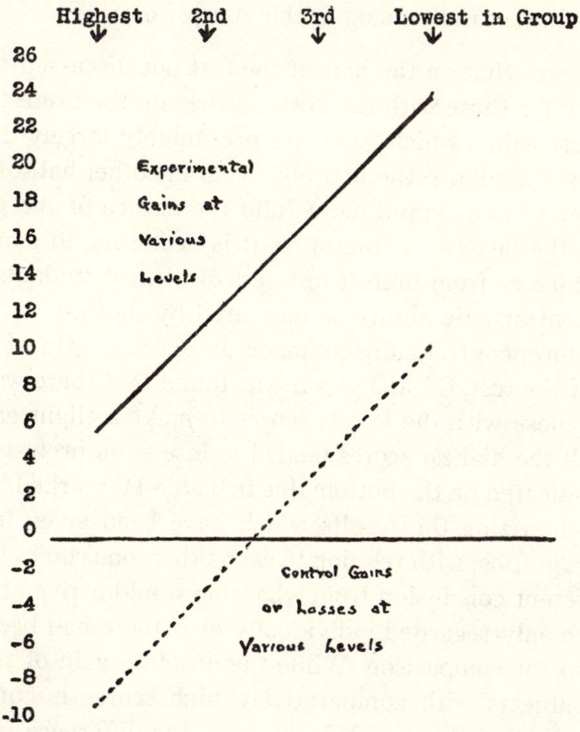

some interesting differences in amount of gain were found. As is shown in Table IV, the average gain of the highest individuals in the groups on the half of the test discussed was 3.1; that of the second highest individuals, 8.2; that of the third highest, 10.3; and that of the lowest, 14.1. This indicates rather definitely that those with the lowest scores tend to gain the most on the half of the test discussed, while those with the highest scores tend to make a gain which is comparatively small.

The middle line in Table IV indicates the average gain at the four levels on the half of the test not discussed by the groups. The average numerical gains on the half not discussed were as follows:

For the highest individuals in the groups, 3.1.
For the second highest individuals in the groups, 3.1.
For the third highest individuals in the groups, 7.5.
For the lowest individuals in the groups, 9.9.

This indicates that on the half of the test not discussed there was a tendency for those with the lowest scores on the pretest to make the greatest gains, which gains are presumably largely due to the discussion of similar esthetic problems in the other half of the test.

In order to understand more fully the nature of the gains discussed in the last two paragraphs, it is necessary to consider the gains and losses from pretest to retest of control students on various levels of artistic ability as measured by the test. As has been stated before, control subjects made an average gain of .5 points on half of the test. On analysis it was found that there was a tendency for those with the lowest scores to make a slight gain, while those with the highest scores tended to lose from pretest to retest. This is indicated on the bottom line in Table IV marked "control."

In summarizing the results which have been given in the last three paragraphs, with relation to each other, one comes to a somewhat different conclusion from what one would arrive at if the results were only regarded individually or if there had been no control group for comparison. While the absolute gain of the experimental subjects with comparatively high scores is considerably less than that of those with low scores, the difference in gain between the individuals at various levels of artistic ability as measured by the art test practically disappears when the control gains and losses for the various levels are taken as a base.* When looked at from this standpoint, those with high scores made as great

* See Table V.

gains from pretest to retest as did those with low scores. However, it should be kept in mind that the average absolute gain of experimental subjects with comparatively low scores was considerably greater than the absolute gain of those with comparatively high scores.

Summary of results pertaining to individual shifts. The following results have been discussed in this chapter:

1. Evidence was given which tended to indicate that individuals in those cultural groups which are most influential in discussion are themselves least influenced by discussion.

2. No significant relationship was found between scholastic ability and influenceability. Nor was there any relationship found between certain personality characteristics studied and influenceability, except that the students most desired by classmates as personal friends are least influenced on the average.

3. There was evidence that students with extreme views on social, educational, economic, and political views are neither significantly more or less influenced by discussion than are the middle-of-the-roaders.

4. Evidence indicates that the discussion of esthetic problems leads students to make individual judgments more similar to those of experts than would be made if the discussion had not taken place.

5. Discussion of esthetic problems also leads students to make esthetic judgments more similar to those of experts on problems similar to those discussed. The gain, however, was not as great here as on the problems discussed.

6. An inverse relationship was found between size of pretest score on the art test and amount of shift from pretest to retest. This was true for both experimental and control subjects.

7. An inverse relationship was found in the experimental group between size of pretest score on the art test and amount of gain from pretest to retest. In the control group those with low scores on the pretest made slight gains on the average, while those with high scores tended to lose slightly from pretest to retest.

CHAPTER VII

IMPLICATIONS OF THE STUDY

THE IMPLICATIONS which are given in this chapter are not to be considered as thoroughly proved but rather are to be thought of as representing some of the tentative beliefs held by the writer. While not all of the implications are directly based on the results which have been discussed in the study, they have grown out of the writer's experience with scores of discussion groups.

Is discussion worth while? Several comments can probably best be given under the general question, "Is discussion of value?" the general answer to which necessarily is, "It all depends . . . "

It depends first of all on the purpose in mind in conducting the discussion. If the discussion has as its purpose the development of followers who uncritically accept what they hear and read, then the purpose will probably not be realized. If the purpose back of a particular discussion is to get the greatest amount of information memorized, the purpose could probably be realized more rapidly by means of books. Again, if the purpose of the group is to exchange and criticize ideas on a certain subject and no one in the group knows much about the subject, then obviously information will not spring full-blown into the minds of the discussers simply because several people are gathered together for discussion, but must be gleaned from other sources outside that particular discussion group.

However, discussion would generally seem to be of considerable value if it has any of the following purposes:

1. To have individuals' ideas on a particular subject exposed and inspected for flaws, since undiscussed information tends to have unsuspected weaknesses in its makeup.
2. To develop critical thinking.
3. To develop the ability to have one's ideas held up to criticism

without such criticism producing an emotional upset in the individual whose ideas are criticized.

4. To develop leaders of discussion.

5. To give those with the most and best information an opportunity to have the greatest weight in formulating the group decision.

6. To emphasize the tentative nature of the weight of the evidence.

7. To develop the ability to explain with clarity one's ideas in a social setting, which is a very different setting in a psychological sense from an isolated setting such as that in which much writing, for instance, is done.

8. To create group constructed ideas and plans which are more significant than the ideas or plans of any single member of the group and even more valuable than any mere *combination* of the ideas held by members of the group at the beginning of the discussion. (The achievement of this purpose represents the results of a most desirable type of discussion, the ideal type in the opinion of some experts in the field.)

It should be constantly borne in mind that whether discussion or some other form of activity can most profitably be engaged in at any particular time depends to a very great degree upon the specific situation, and particularly on the following factors and their relationships to each other in that specific situation:

1. Amount of knowledge held by prospective discussion participants.

2. The type of situation in which the information discussed is later to be used. Generally speaking, information should be learned, if possible, in the way it is later to be used.

3. The amount of time available.

4. The size of the group.

5. The form of the material. Lack of preparation of materials for discussion is likely to produce as undesirable results as lack of preparation in other ways.

6. The amount of training, direct or indirect, in discussion which prospective discussers have had.

Can the technique of discussion be taught? The writer believes that the ability to influence others in discussion, regardless of the type of influence desired, is largely an acquired ability, assuming that the individual has a level of mental ability which approximates that possessed by others to be influenced in a particular situation. However, before the art of discussion can be adequately taught from any of the several angles previously suggested under purposes of discussion, much investigation and study need to be done. Also, there is need for greater recognition of the fact that training for discussion involves not a single factor, but rather includes a variety of factors, such as training in the ability to swing a group decision, training so that much information may be gleaned from discussion, training for critical or analytical purposes, or training in the ability to aid the group create a new and better plan than any that has yet been proposed.

Are pencil and paper tests of dominance valid? This study certainly does not give any conclusive answer on this point. The results are fairly clear-cut, but the interpretation of these results is open to question. The fact is that no significant relationship was established between dominance as measured by the Bernreuter scale and discussion dominance as measured by the influence scales. This may be due to either: (1) errors in the measurement of discussion dominance, (2) lack of validity of the measure of general dominance, (3) lack of relationship between general dominance and this specific type of dominance, or (4) a combination of all these three. The writer is at present inclined to believe that "dominance" is very much a part of the specific situation in which the individual finds himself, and that while there is probably a "dominance core" exhibiting itself to some degree in all social situations, it is doubtful whether a general dominance scale will be of great value in indicating how an individual is likely to behave in a specific situation or a particular kind of social setting.

Marks and influence. Marks and the usefulness of the marking system have been seriously questioned for some time and probably justly so. In this study a fairly high correlation (.38) was found between marks and ability to influence others in discussion. While, of course, this should not be given too great weight in determining whether marks shall be retained, it does appear that they do have a small amount of value as indicators of ability along the line investigated in this study.

The subtlety of influence of the types studied in this investigation. The writer's experience in conducting this study leads him to believe that influence is a very subtle thing. In other words, individuals who engage in discussion frequently feel that although others have probably been influenced by discussion, they themselves have been relatively impervious to the expressed opinions and ideas of others, while as a matter of fact they themselves have probably had their own ideas given a rather healthy shake-up and they no longer see things as they did before discussion. An incident which actually occurred during the experiment will serve to illustrate this point. The writer was particularly interested to find from pretest to retest on the art test the possible change of opinion of a student who talked a great deal in her group and really felt she knew a good deal about art—probably more than the others in her group. As this student handed in her retest paper the writer casually asked her whether she thought her answers were much different from those she had given the first time she had taken the test. "No," she replied quickly and rather positively, "I disagreed with the other three girls on practically every question discussed and still disagree." As a matter of fact, when her paper was scored it was found that she had gained 17 points from pretest to retest—apparently without realizing any particular changes in her responses. Influence seems to be exerted inevitably and frequently in a very subtle fashion when two or more are gathered together at the discussion table. (No attempt has been made in this study to analyze the methods or techniques used in influencing others, a subject studied in some detail by Bonney [1] although without particular reference to discussion.)

CHAPTER VIII

SUMMARY AND CONCLUSIONS

The problem

The major problem of this study was to ascertain the individual traits and background characteristics related to influence in discussion and those related to being influenced by discussion. Influence in discussion, as used in this study, refers both to getting others to adopt one's point of view or something similar to it as the group decision, and to affecting others so that in their individual decisions made subsequently to the discussion they will significantly reflect the point of view of the one who exerted the influence in discussion. The first type of influence has been termed immediate group influence; the latter type has been called persisting individual influence. Influenceability is the term which has been used to indicate the amount an individual changed his responses from pretest to retest with discussion intervening. A comparatively minor issue of the study was to determine whether discussion of esthetic material tended to lead students to make esthetic judgments more or less similar to those of experts.

Subjects

The subjects of the study were 185 college women, largely sophomores and juniors, taking a course in either Introductory Psychology or Educational Psychology (with 4 exceptions) at the time of experimentation. Of the total number of subjects, 161 served as experimental subjects and 24 as control subjects. The subjects were not informed that an experiment was being conducted, lest such knowledge might tend to vitiate the results.

Subject matter used for discussion

In the experimental discussions, eight series of problems were used. Seven of these series, which involved decisions on current controversial educational, economic, political, and social issues, were constructed by the writer. The eighth series involved decisions on the McAdory Art Test. Each of the problems in the series constructed by the writer was arranged with three major alternative answers placed on a linear scale. The scale had one major alternative at each end and a third major alternative in the center of the response device. Actually, as the scale was designed for each problem, there were an almost unlimited number of minor alternatives, any one of which might be designated as the best solution to a particular problem.

In the experimental technique developed by the writer for this study, two types of individual influencing ability were measured: immediate group influence and persisting individual influence. Immediate group influence refers to the relative amount of influence an individual had in formulating the group decision which was made immediately after the discussion of a particular problem. Persisting individual influence is the term used to describe the amount of effect an individual had on the retest responses of those who had previously discussed with her. To get these two measures it was necessary to have the following information for each discussion group, of which there were 156:

1. The position of each individual before discussion on every problem to be discussed by the group. This was determined by using a pretest five days before the discussion of each series.

2. The mean position of the group before discussion on each problem. This was found by using the individual pretest positions of the members of the group.

3. The group decision on each problem discussed. This was indicated immediately after the discussion of each problem.

4. The position of each member of the group after discussion of each problem. This was obtained by a retest five days after discussion of each series.

5. To get fairly reliable measures of these influencing abilities, all experimental subjects, with a few unavoidable exceptions due to absences, engaged in four fifty-minute supervised discussions which were held during the course of a school semester about three weeks apart.

MEASURING INSTRUMENTS USED AND OTHER DATA GATHERED

1. The Otis Test of Mental Ability.

2. The McCall Multi-Mental Test.

3. The Scholastic Aptitude Test of Verbal Ability.

4. The Scholastic Aptitude Test of Mathematical Ability.

5. Teachers' marks for the semester in which the discussions took place.

6. Bernreuter Personality scales which give measures of: Emotional stability, self-sufficiency, introversion-extroversion, dominance-submission.

7. Flanagan scales based on the Bernreuter Inventory which give measures of: Self-confidence, and sociability.

8. Maslow Social Personality Inventory for Women. A test of dominance feeling.

9. Self-rating sheet for dominance feeling.

10. Classmates' ratings for likability.

11. Hartmann Social Attitudes Inquiry, a test of radical tendencies.

12. Suggestibility and influenceability evaluations.

13. Religious background.

14. Year in school.

15. Type of elementary school attended.

16. Number of children in family.

17. Position in family.

18. McAdory Art Test.

FINDINGS CONCERNED WITH THE RELATION OF INFLUENCE IN DISCUSSION TO MEASURES OF MENTAL ABILITY

1. No significant relationship was found between either type of influence, immediate group or persisting individual, on the one hand, and mental ability as measured by either the Otis Test of Mental Ability or the Multi-Mental Test, on the other hand. However, the validity of each of these tests as measures of mental ability of college students is seriously questioned on the basis of

the almost complete lack of correlation between either of these tests and college marks.

2. Statistically reliable positive relationships were found between scores on the Scholastic Aptitude Test of Verbal Ability and both types of influence in discussion ability, immediate group and persisting individual influence. In other words, on the average, those who are most influential in discussion tend to make high scores on this Scholastic Aptitude Test.

3. The correlation between immediate group influence and mathematical ability as measured by the Scholastic Aptitude Test of Mathematical Ability was found to be .22 ± .066, indicating that the chances are about 99 in 100 that a positive relationship really exists between these two variables in the populations from which our sample populations were drawn. A statistically reliable positive relationship was found between persisting individual influence and mathematical ability as measured by the Scholastic Aptitude Test, the actual coefficient of correlation being .36 ± .066.

4. When average semester marks were correlated with immediate group influence, a statistically reliable coefficient of correlation of .38 ± .05 resulted, indicating that, in general, students able to influence their classmates most in discussion situations were also able to influence their teachers most favorably and get the highest marks. Class marks also correlated positively with persisting influence although not to as great a degree as they correlated with immediate group influence.

Findings related to personality indices

1. No statistically reliable relationship was found between either (*a*) emotional stability, or (*b*) self-sufficiency, or (*c*) introversion-extroversion, or (*d*) dominance as measured by the Bernreuter pencil and paper scales on the one hand, and either (*a*) immediate group influence, or (*b*) persisting individual influence on the other. The lack of any apparent relationship between the pencil-and-paper dominance test and dominance as measured in an actual face-to-face situation is particularly interesting.

2. A small amount of relationship, bordering on statistical re-

liability, was found between self-confidence as measured by the Flanagan scale and immediate group influence. No reliable degree of relationship was found between self-confidence and persisting individual influence.

3. No relationship was established between sociability as measured by the Flanagan scale, and either immediate group influence or persisting individual influence.

4. When the Maslow dominance results were correlated with influence scores of the two types, very low positive coefficients of correlation resulted. However, in neither case was any statistically reliable relationship found.

5. Scores based on a self-rating of "dominance feeling" produced no reliable coefficient either when correlated with immediate group influence or when correlated with persisting individual influence.

6. Scores on the Social Attitudes Inventory, a radicalism test, when correlated with immediate group influence scores yielded a coefficient which, with its probable error, indicated that the chances are about 99 in 100 that a positive relationship exists between radicalism and immediate group influence. When radicalism scores were correlated with persisting individual influence scores, a positive coefficient again appeared, although it was not statistically reliable.

7. Religious background was found to be a factor significantly related to influence in discussion. Jewish students were found to be, on the average, most influential in discussion. With regard to immediate group influence, Protestants followed Jewesses in influence, with Catholics placing last. The difference between the average immediate influence score in the Jewish group and the mean score in the Catholic group was statistically reliable; that between Jewesses and Protestants fairly reliable; and that between Protestants and Catholics not reliable. With respect to persisting individual influence, the Jewish group again ranked first, with Protestants and Catholics having approximately the same average score. The difference in persisting individual influence scores favoring the Jewish group as compared with the other two groups was not statistically significant.

Chief findings relating to amount and kind of individual shifts

1. There is a tendency for those who are most influential in discussion to be themselves least influenced by the discussion.

2. Influenceability is apparently not significantly related to either scholastic ability, the personality characteristics investigated (except in the case of desirability as a friend where an inverse relationship was found to exist), or extremeness of views on questions discussed.

3. Discussion of esthetic problems tends to "improve" the judgments of students on problems discussed, and also on similar esthetic problems.

4. Those who made the best scores on the art test made fewer shifts and smaller gains from pretest to retest than those who made comparatively low scores on the test. This was found to be true for both the experimental and the control group, with those in the control group who made the highest scores on the pretest tending to register small losses from pretest to retest.

BIBLIOGRAPHY

BOOKS

1. Bonney, M. E. *Techniques of Social Control.* George Banta Publishing Co., Menasha, Wis., 1934.
2. Bowman, Le Roy C. *How to Lead Discussion.* The Womans Press, New York, 1934.
3. Busch, H. M. *Leadership in Group Work.* Association Press, New York, 1934.
4. Chen, W. K. *The Influence of Oral Propaganda Material upon Students Attitudes.* Archives of Psychology. Columbia University, 1933.
5. Coyle, G. L. (Editor). *Studies in Group Behavior.* Association Press, New York, 1937.
6. Elliott, Harrison S. *The Process of Group Thinking.* Association Press, New York, 1928.
7. Fansler, Thomas. *Discussion Methods for Adult Groups.* American Association for Adult Education, New York, 1934.
8. Follett, M. P. *The New State: Group Organization the Solution of Popular Government.* Longmans, Green and Co., New York, 1926.
9. Follett, M. P. *Creative Experience.* Longmans, Green and Co., New York, 1924.
10. Leigh, R. D. *Group Leadership.* W. W. Norton and Co., New York, 1936.
11. Lindeman, E. C. *Social Education.* New Republic, Inc., New York, 1933.
12. Murchison, Carl. *Handbook of Social Psychology.* Clark University Press, Worcester, Mass., 1935.
13. Sherif, M. *The Psychology of Social Norms.* Harper and Brothers, New York, 1936.
14. Sheffield, Alfred D. *Creative Discussion.* The Inquiry, New York, 1927.
15. Sheffield, Alfred D. *Training for Group Experience.* The Inquiry, New York, 1929.
16. Slavson, S. R. *Creative Group Education.* Association Press, New York, 1937.
17. Studebaker, J. W. *Plain Talk.* The National Home Library Foundation, Washington, D. C., 1936.
18. Tead, Ordway. *The Art of Leadership.* McGraw Book Co., New York, 1935.

19. Walser, Frank. *The Art of Conference.* Harper and Brothers, New York, 1933.
20. Znaniecki, Florian. *Social Actions.* Farrar and Rinehart, New York, 1936.

TESTS AND RELATED MATERIALS

21. Allport, Gordon W. and Floyd, H. "A Test for Ascendence-Submission." *Journal of Abnormal and Social Psychology.* Vol. 23, 2. 1928.
22. Bernreuter, Robert G. *The Personality Inventory.* Stanford University Press. Stanford, California. 1935.
23. Bernreuter, Robert G. *Manual for the Personality Inventory.* Stanford University Press. Stanford University, California. 1935.
24. College Entrance Examination Board. *Thirty-sixth Annual Report of the Secretary.* New York. 1936.
25. Eisenberg, Philip. "Expressive Movements Related to Feeling of Dominance." *Archives of Psychology.* No. 211. New York. 1937.
26. McAdory, Margaret. *McAdory Art Test.* Bureau of Publications, Teachers College, Columbia University. New York. 1929.
27. McCall, William. *McCall Multi-Mental Test.* Bureau of Publications, Teachers College, Columbia University. New York. 1930.
28. Otis, Arthur S. *Otis Self-Administering Tests of Mental Ability.* World Book Co., Yonkers, New York. 1922.
29. Siceloff, M. M. and Woodyard, Ella. *Validity and Standardization of the McAdory Art Test.* Bureau of Publications, Teachers College, Columbia University. New York. 1933.
30. The John Dewey Society. *The Teacher and Society.* D. Appleton-Century Co. New York. 1937.

VITA

Ray Hamill Simpson was born in Indiana County, Pennsylvania, September 2, 1907.

Academic training: Graduated from Indiana High School (Pa.), 1925. B. S. in Education, Indiana State Teachers College (Pa.), 1930; M. A., Teachers College, Columbia University, 1934.

Professional Experience: Teacher of rural elementary school, 1925-1926; teacher in Junior High School, Indiana, Pa., 1929-1930; teacher in Senior High School, Punxsutawney, Pa., 1930-1934; Lecturer in Psychology, Barnard College, 1934-1936; Assistant in Educational Psychology, Teachers College, Columbia University, Summer Sessions of 1934 and 1935; Instructor in Psychology, Barnard College, since 1936.